THE POWER OF LIMITS

The Power of Limits

Proportional Harmonies in Nature, Art, and Architecture

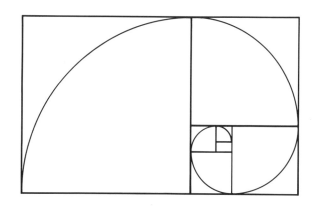

György Doczi

 1994 *Shambhala*   *Boston & London*

SHAMBHALA PUBLICATIONS, INC.
Horticultural Hall
300 Massachusetts Avenue
Boston, Massachusetts 02115
www.shambhala.com

Distributed in the United States by Random House and in
Canada by Random House of Canada Ltd
Printed in Canada
♾ This edition is printed on acid-free paper that meets the
American National Standards Institute Z39.48 Standard.

9   8   7   6

Library of Congress Cataloging in Publication Data
Doczi, György, 1909–
        The power of limits.

        Bibliography: p.
1. Creation (Literary, artistic etc.)   2. Aesthetics.
3. Proportion.   4. Nature (Aesthetics)
I. Title.
BH301.C84D62   700′.1   77-90883
ISBN 0-87773-193-4 (pbk.)

Copy editor: Holland D. Hammond; designer: Hazel Berchoiz;
production manager: James M. Gimian; production artist:
Bea Ferrigno-Lee; paste-up: Liza Matthews

Contents

In memory of Sven Ivar Lind

Preface

Why do apple blossoms always have five petals? Only children ask such questions. Adults pay little attention to such things, taking them for granted, like the fact that we use only as many numbers as we can count on our ten fingers. When we look deeply into the patterns of an apple blossom, a seashell, or a swinging pendulum, however, we discover a perfection, an incredible order, that awakens in us a sense of awe that we knew as children. Something reveals itself that is infinitely greater than we are and yet part of us; the limitless emerges from limits.

This book searches for some of the basic pattern-forming processes that, operating within strict limits, create limitless varieties of shapes and harmonies. It is an interdisciplinary venture into the no-man's-land between the borders of science, art, philosophy, and religion, an area that has been largely disregarded in recent years because its contents are intangible. This area, however, bears investigation, since the powers that shape our lives and our values have their source here.

As René Dubos points out in *So Human An Animal,* this age of affluence and technological achievement is also an age of anxiety and despair. Traditional social and religious values have eroded to the point where life often seems to have lost its meaning. Why isn't the harmony that is apparent in natural forms a more powerful force in our social forms? Perhaps it is because, in our fascination with our powers of invention and achievement, we have lost sight of the power of limits. Yet now we are forced to confront the limits of the earth's resources, and the need to limit over-population, big government, big business, and big labor. In all realms of our experience, we are finding the need to rediscover proper proportions. The proportions of nature, art, and architecture can help us in this effort, for these proportions are shared limitations that create harmonious relationships out of differences. Thus they teach us that limitations are not just restrictive, but they also are creative.

It is not an accident that an architect should be writing such a book, for it is the business of architects to work with proportions. This architect is old. It took him a lifetime to attempt to answer the questions he asked as a child. These answers may not satisfy the experts, and they won't still the curiosity of a single child, but they might lead to further, perhaps more fruitful questions about the puzzles and beauties hidden in the patterns and proportions of this world.

György Doczi
Seattle, Washington

Acknowledgments

This book could not have been completed without the patient—and at times not-so-patient—but always faithfully sustained support of my wife. Many others also helped me: chief among these were my brother and daughter, as well as members of the Art and Music Department of the Seattle Public Library, Gerald Dotson, Marilyn West, Regina Hugo, David and Miriam Yost, David Tomlinson, Dr. Werner and Margit Weingarten, Dr. Richard M. Braun, Donald Collins, Brian Brewer, Rabbi Joseph Samuels, John A. Sanford, John Fuller and the staff of Shambhala Publications, headed by Samuel Bercholz.

Research for this book was greatly benefited by access to collections in various departments of the University of Washington, Seattle, granted by Dr. Daniel O. Graney in Biological Structures and by Dr. John Edwards in Entomology, and by access to the collection at the Thomas Burke Memorial Museum of the University, granted by Don Coburn, senior restorer of prehistoric skeletons, by Bill Holm, curator of Northwest Coast Indian Art, and by assistant director Robert Free.

My study of aquatic animals at the Seattle Aquarium was greatly facilitated by general curator Dr. John W. Nightingale, and my study of fishes of Canadian Pacific waters was aided by the courtesy of the Canadian Minister of Supply and Services. Professor Donald J. Borror, his co-authors, and their publishers helped by permitting me to use their illustrations in *A Field Guide to the Insects,* Houghton and Mifflin Co., and *An Introduction to Insects,* Holt, Rinehart and Winston, as models for my own proportional drawings. Tom Cole of the Boeing Company's public relations department provided me with dimensioned drawings of the 747. To all of these, and to many other helpers—here unnamed—my sincere gratitude. For errors and mistakes which, in spite of all this good help, might have crept into the book, I alone am responsible.

Fig. 1. Center of a daisy.

CHAPTER 1: Dinergy in Plants

It is said that the Buddha once gave a sermon without saying a word; he merely held up a flower to his listeners. This was the famous "Flower Sermon," a sermon in the language of patterns, the silent language of flowers. What does the pattern of a flower speak about?

If we look closely at a flower, and likewise at other natural and man-made creations, we find a unity and an order common to all of them. This order can be seen in certain proportions which appear again and again, and also in the similarly dynamic way all things grow or are made—by a union of complementary opposites.

The discipline inherent in the proportions and patterns of natural phenomena, and manifest in the most ageless and harmonious works of man, are evidence of the relatedness of all things. It is through the limits of discipline that we can glimpse and take part in the harmony of the cosmos—both in the physical world and in our way of life. Perhaps the message of the Flower Sermon had to do with how the living patterns of the flower mirror truths relevant to all forms of life.

Take for instance a daisy. (fig. 1) The pattern at its center is shown in figure 2. The florets that make up this pattern—here represented by circles—grow at the meeting points of two sets of spirals, which move in opposite directions, one clockwise, the other counterclockwise. (See diagram at center.)

Here two of the spirals have been reconstructed with the help of a series of concentric circles, at distances growing along a logarithmic scale, and a series of straight lines radiating from the center. If we connect the consecutive meeting points of these two sets of opposing lines, we can see the daisy's growth spirals. These spirals are logarithmic and also equiangular, since the angle they describe with the radii remains always the same. This is illustrated by the diagram at the right, which shows that segments representing consecutive stages of growth can be rotated around the center until they completely overlap, like a folded fan, proving that old and new stages of growth all share the same angles and the same proportion.

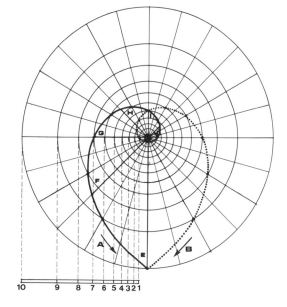

Fig. 2. Diagrams of a daisy. The generating spirals move in opposite directions, are logarithmic, *center,* and equiangular, *right.*

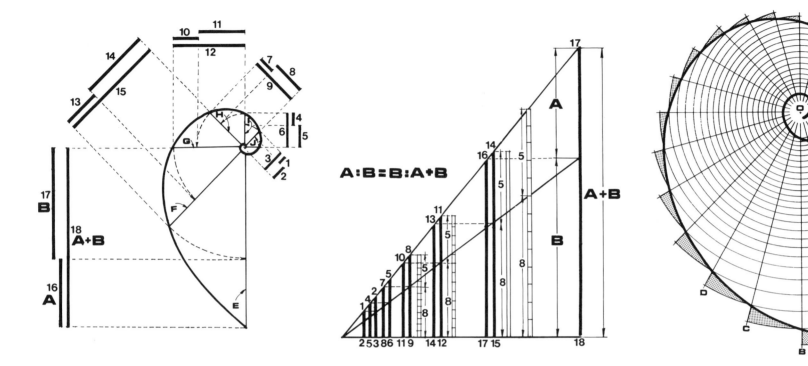

Fig. 3. The golden section in one of the daisy's spirals. Each stage of growth shares the same proportions (see shaded triangles at right).

Just what this proportion is can be seen from figure 3, showing the unfolding of one of these spirals (diagram at left). As the spiral unfolds from the daisy's center, the order of growth, measured along equidistant radii marked **E,F,G,H,I,J,** increases at the same rate. This can be seen from the triangular diagram, where parts of old and new growth increments of each stage, marked in the left diagram by bars, consecutive numbers and also **A's** and **B's,** have been lined up along vertical lines. All points where **A's** and **B's** meet in the triangular diagram fall upon the same slanting line, which meets the vertical scales at distances measuring 5 and 8 respectively. The ratios between these two numbers approximate some remarkably reciprocal features: 5 divided by 8 approximates 0.6 (0.625); 8 divided by 5+8 or 13 also approximates 0.6 (0.615). Conversely, 8 divided by 5 is 1.6, and 13 divided by 8 again approximates 1.6 (1.625), the latter two ratios being the same as the former ones with 1 or unity added.

Expressed in equation form: A:B = B:(A+B). This is the formula of the celebrated *golden section,* a uniquely reciprocal relationship between two unequal parts of a whole, in which *the small part stands in the same proportion to the large part as the large part stands to the whole.*

The term *golden section* derives from the uniqueness as well as the distinctive value attributed to this proportional relationship. On any given line there is only *one* point that will bisect it into two unequal parts in this uniquely reciprocal fashion, and this one point is called the point of golden section. The complete reciprocity of this proportion strikes us as particularly harmonious and pleasing, a fact that has been proven by many scientific experiments since the end of

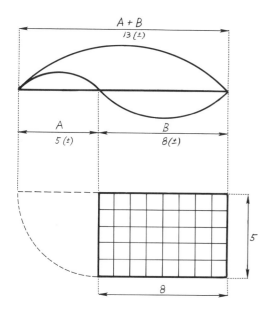

$$A : B = B : (A+B) = 0.618...\qquad B : A = (A+B) : B = 1.618...$$

$$5 : 8 = 0.625; \quad 8 : 13 = 0.615 \qquad 8 : 5 = 1.6; \quad 13 : 8 = 1.62$$

Fig. 4. Approximation of a golden rectangle (5 ÷ 8).

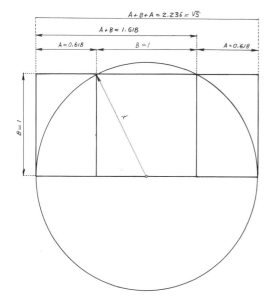

Fig. 5. Classical construction of the golden section, with square within semicircle. Rectangles 1 × 0.618 and 1 × 1.618 are reciprocal golden rectangles.

the last century.[1] Preference for this proportion is also apparent in the tendency of paper standards to approximate it or its multiples, including the sizes of paper money, checks, and credit cards.[2]

Figure 4 shows a so-called *golden rectangle* of 5×8 proportions; it also shows a line bisected by the golden section into parts A=5 and B=8, with similar arcs above and below the line emphasizing the reciprocity of these relationships.

This reciprocity is also illustrated by one of the classical constructions of the golden section, with the help of a square inscribed into a semicircle (fig. 5). A circle drawn from the center of the square's base touching the opposite corners of the square (radius *r*) will produce the golden section's proportions along both sides of the extended baseline.

If the square's sides are 1 unit long, then each of the extensions will be 0.618 unit long and the 1 × 0.618 rectangles on either side of the square will be golden rectangles. Each of these combined with the square form a larger golden rectangle, 1 × 1.618. These larger rectangles and the smaller ones are each other's reciprocals, in the sense that the larger side of the small ones and the smaller side of the large ones are the same. The total length of these reciprocal golden rectangles is 2.236 units, this number being identical with $\sqrt{5}$.

It has often been demonstrated that the golden section's proportions are frequent in patterns of organic growth, particularly between neighboring old and new increments. This is the reason why the biologist C. H. Waddington proposed to call this proportion the *relatedness of neighbors*.[3]

Patterns generated by spirals moving in opposite directions are frequent in nature, as we shall see. Here they concern us as special instances of a more general pattern-forming process: the union of complementary opposites. Sun and moon, male and female, positive and negative electricity, Yin and Yang—the union of opposites has been since ancient times an important concept in mythologies and mystery religions. The two parts of the golden section's proportions are unequal: one is smaller, the other larger. They are often referred to as *minor* and *major*. Minor and major here are opposites united in a harmonious proportion. The process itself by which the daisy's harmonious pattern was reconstructed is likewise a joining of complementary opposites—straight radii and rotating circles.

Many terms refer to aspects of the pattern-forming process of the union of opposites, but strangely enough none expresses its *generative* power. *Polarity* refers to opposites but without the indication that something new is being born. *Duality* and *dichotomy* indicate division, but do not mean joining. *Synergy* indicates joining and cooperation, but does not refer specifically to opposites.

Since there is no adequate single word for this universal pattern-creating process, a new word, *dinergy*, is proposed. *Dinergy* is made up of two Greek words: *dia*—"across, through, opposite;" and "energy." In the daisy this dinergic energy is the creative energy of organic growth. To what extent other plants show dinergic pattern formations shall be our next concern.

Windows on the infinite

The center of the sunflower (fig. 6) is also composed of florets, which eventually turn to seeds, and they also grow along logarithmic, equiangular spirals, moving in opposite directions, which are joined dinergically as in the daisy. In the drawing of the model small pyramids approximate the shapes of the seeds. The spirals which are formed by this seed pattern are detailed in figure 7. Diagram **a** shows that spirals **A** and **B** are the outlines of the seeds and **C** and **D** are their diagonals.

If one follows the different curvatures of these spirals through the squares formed by the radiating and rotating sets of lines in diagram **a,** one can see that spiral **A** moves from one circle to the next as well as from one radius to the next within a single row of squares. We will call this a curvature of 1:1, signifying that the rotational and the radiational components of growth are equal. Spiral **B** moves through two squares while reaching from one radius to the next, crossing two circles: a curvature of 1:2. In a similar manner one can say that spiral **C** has a curvature of 3:1, while spiral **D** approximates 5:1.

In spite of their differences in curvature, all of these spirals share the qualities of being logarithmic and equiangular, through all stages of growth. Such stages are shown in diagram **b** marked upon each spiral by lower-case letters. The spiral diagram **c** at the right shows, on the example of spiral type **C,** how succeeding segments of growth, which follow each other in cyclic order, can be folded upon each other fan-like, as in the daisy. This demonstrates that they are of the same shape even though different in size, sharing the same proportional rate of growth. That this rate is again the rate of the golden section is shown by diagram **d.**

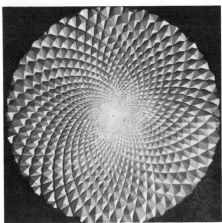

Fig. 6. Sunflower and model of seed pattern.

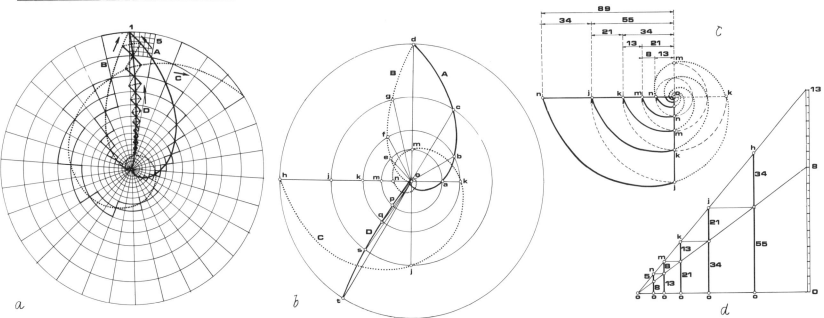

Fig. 7. Typical spirals of the sunflower's seed pattern.

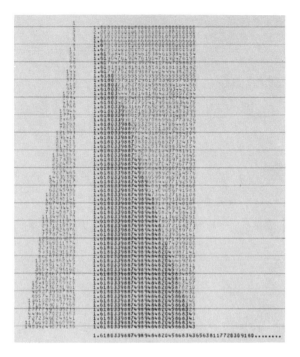

Fig. 8. Computer print-out of golden section's
ϕ ratio and Fibonacci series.

The numbers signifying neighboring old and new stages of growth prove to be members of a so-called *summation series,* in which each number is the sum of the two previous ones: 1, 2, 3, 5, 8, 13, 21, 34, 55, 89, 144, 233, 377, etc. This is the famous *Fibonacci series,* after the nickname of Leonardo of Pisa who introduced it into Europe about eight hundred years ago, together with Hindu-Arabic numerals and the decimal system. Any number in this series divided by the following one approximates 0.618 . . . and any number divided by the previous one approximates 1.618 . . . , these being the characteristic proportional rates between minor and major parts of the golden section. In literature this rate is frequently referred to with the Greek letter phi (ϕ).

The three dots after the numbers indicate that these numbers are "irrational," so called because they can only be approximated, never expressed fully in a fraction. One can keep dividing any of the neighboring numbers—after 13—without ever reaching an end. An IBM computer once produced 4000 digits of this number at which point it was stopped without reaching a rational number. The computer printout in figure 8 was stopped at the forty-fifth place after the decimal point.

It has been recorded that the Pythagoreans of ancient Greece, who are credited with having discovered in the sixth century B.C. the infinite nature of irrational numbers, were puzzled, frightened and filled with awe upon their discovery and tried to keep it a secret, proscribing the death penalty for those who would dare to divulge it. Legend has it that a violator of this prohibition, who escaped to sea, drowned there. His death was attributed to divine punishment.

The numbers of the Fibonacci series curiously reappear in the total number of the sunflower's spirals. The model in figure 6 was made from a sunflower head that had 34 and 55 spirals; 34:55 = 0.6181818 . . . . Sunflowers with 89 and 144, and with 144 and 233 opposite spirals have also been reported; 89:144 = 0.6180555 . . . ; 144:233 = 0.6180257 . . . .

One does not have to fear vengeful deities to feel something like awe at such unexpected precision in a pattern of natural growth. It seems unreasonable to believe that the number of seeds in a sunflower is preordained, yet something like that is exactly what happens. Irrational numbers are not unreasonable; they are only *beyond* reason, in the sense that they are beyond the grasp of whole numbers. They are infinite and intangible. In patterns of organic growth the irrational ϕ ratio of the golden section reveals that there is indeed an infinite and intangible side to our world.

> *To see a world in a grain of sand*
> *and heaven in a wild flower,*
> *hold infinity in the palm of your hand*
> *and eternity in an hour.* [4]
>   —William Blake

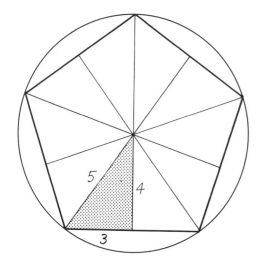

Fig. 10. Apple blossoms, apples and pears, and flower of loganberry bush.

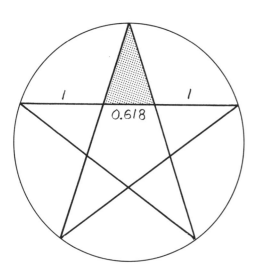

Fig. 9. Pentagon and pentagram showing Pythagorean triangle's and golden section's proportions.

Every daisy and sunflower is a window on the infinite, as are apple blossoms and the flowers of other trees and bushes bearing edible fruits. These grow according to the pattern of the pentagon and its extension, the pentagonal star or pentagram, (fig. 9) in which neighboring lines relate to each other in the dinergic golden relations of neighbors. Apples and pears, when cut through their girth, reveal the pentagonal star pattern in their seed structure, inherited from their original flower pattern. (fig. 10)

Each of the triangles in the pentagonal star have two equal sides that relate to the third side as 8 does to 5, or as 1.618 relates to 1 . . . . These reciprocal relationships can be seen when the pentagram is combined with the golden section's construction, as in figure 11, creating a $\sqrt{5}$ rectangle, consisting of reciprocal golden rectangles. The smaller rectangle's sides are identical with the sides of the pentagram's triangle.

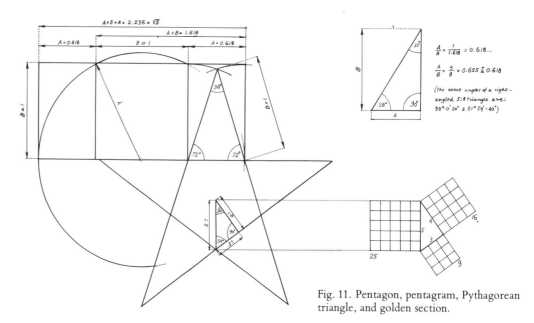

Fig. 11. Pentagon, pentagram, Pythagorean triangle, and golden section.

The sides of the right-angled triangle, ten of which make up the pentagon, also approximate the dinergic golden relations of neighbors. As shown within the pentagonal star, these sides approximate 3-, 4-, and 5-unit lengths, and 3 and 5 are neighboring members of the Fibonacci series (3:5 = 0.6). This 3–4–5 triangle is at times referred to as the *Pythagorean triangle,* because it illustrates the Pythagorean theorem (the square of the hypotenuse of a right triangle is equal to the sum of the squares of the other two sides). This triangle is frequently found in plant patterns, as figure 12 shows.

The pentagonal star was the sacred emblem of the Pythagorean fellowship, which consisted of men and women living in communities and abstaining from all luxuries, dedicated to a life of moderation and the practice of healing. The Pythagoreans applied the letters of the name "Hygeia," the goddess of healing, to the tips of their sacred emblem. The five-pointed star has remained a universal symbol of good portent, appearing in the national flags of many countries.[5]

The pentagon and pentagram, like all patterns, are defined by their limits. Incorporated in the harmonious patterns of fruits and flowers, they exemplify an epigram attributed to Pythagoras, that *limit gives form to the limitless.* This is the power of limits.

Fig. 12. Pythagorean 3-4-5 triangle in plants.

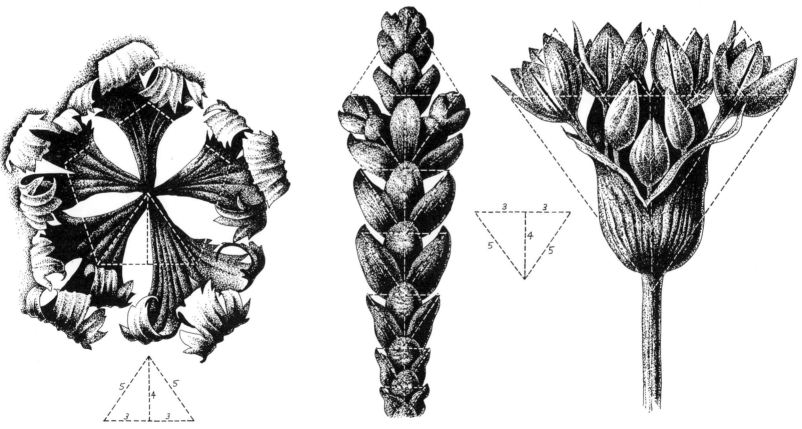

Dry leaf of Globe Flower (Trollius Europeus)   Deerhorn Cedar (Thujopsis dolabrata)   Garlic. (Allium Ostrowskianum.)

Harmonies of music and growth

By *harmony* we generally mean a fitting, orderly and pleasant joining of diversities, which in themselves may harbor many contrasts. In this sense, harmony is a dinergic relationship, in which different and often contrasting elements complement each other by joining. That such dinergic joining is at the heart of all harmonies is suggested by the origin of the word *harmony,* from the Greek *harmos,* to join.

The concept of harmony also goes back to Pythagoras, who according to legend discovered it while listening to the sound of hammering that came from different anvils in a smith's shop. This observation guided him by analogy to other instruments, such as the vibrating strings of a lyre. He found that two strings sound most pleasant together when they are equal, or when one is plucked at 1/2, 2/3, or 3/4 of the other's length; in other words, when the length of the plucked strings relate in proportions expressible in the smallest whole numbers: 1, 2, 3, 4. (fig. 13)

The **1:1** proportion, which is identity, is called *unison.* The **1:2** proportion, which produces the same sound as the full string, only at a higher pitch, is called the *octave* because it reaches through all eight intervals of the scale (the eight white keys of the keyboard). The Greeks called this proportion *diapason: dia,* "through," *pason,* from *pas* or *pan* meaning "all." The pleasant sound of the **2:3** proportion was called *diapente* (*penta,* "five"), today called the *fifth,* reaching through five intervals. The consonance of the **3:4** proportion was called *diatessaron* (*tessares,* "four"), or the *fourth.*

The 2:3 = 0.666 proportion of diapente is a close approximation of the golden section's 0.618 . . . ratio. Diatessaron is identical with the 3:4 proportion of the Pythagorean triangle. Diapason, the octave, has the 1:2 = 0.5 proportion of a rectangle composed of two equal squares, having a diagonal of $\sqrt{5}$ length, which is the joint length of two reciprocal golden rectangles. (fig. 14)

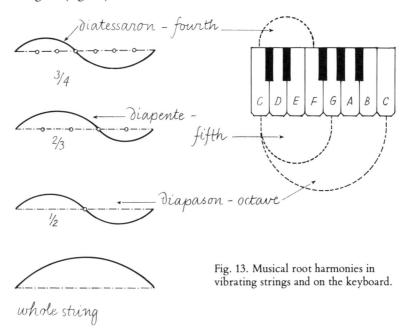

Fig. 13. Musical root harmonies in vibrating strings and on the keyboard.

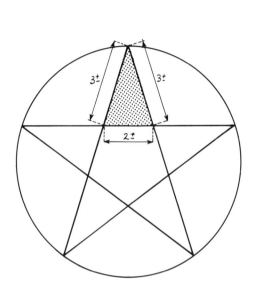

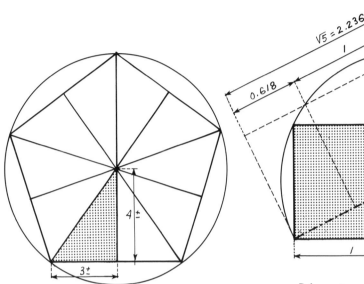

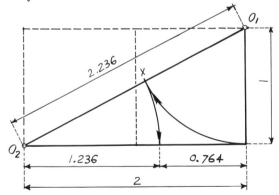

Diapente – or 2:3 – corresponds to the sides of the pentagram's triangle, approximated by 2:3 = 0.666... $\overset{+}{=}$ 0.618 = ϕ

Diatessaron – or 3:4 – corresponds to the pentagon's triangle, approximated by a 3-4-5 triangle. 3:4 = 0.75

Diapason – or 1:2 – corresponds to a rectangle composed of two squares, with a diagonal of 2.236 = = $\sqrt{5}$, which is the side length of two reciprocal golden rectangles – or a square plus two lateral golden rectangles.

One way of constructing the golden section's proportion along a straight line (see diagram at the right) follows diapason harmony. — A circle from O_1 center with a radius of 1 establishes point X on diagonal O_1 - O_2. A second circle with O_2 as center and O_2 - X radius bisects the base of the 2 x 1 rectangle in golden section. 0.764 : 1.236 = 0.618 / 229 = ϕ

Fig. 14. Visual equivalents of Pythagorean musical harmonies and construction of the golden section.

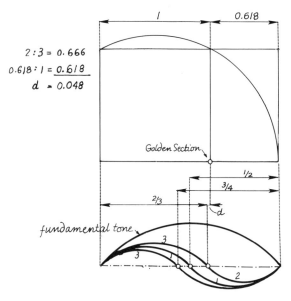

$2:3 = 0.666$
$0.618:1 = 0.618$
$d = 0.048$

Fig. 15 Harmonic overtones of vibrating string and the golden section.

Pursuing further the role of dinergy in musical harmonies, we find the 1:2, 2:3 and 3:4 proportions reappearing in the first and strongest overtones, also called partials or harmonics, which reverberate within every musical sound, blending with its fundamental tone, as if additional, invisible strings were being plucked simultaneously, accompanying and complementing the fundamental. It is this dinergic joining of harmonics with the fundamental that gives musical sounds their fullness, vitality and beauty—referred to as *timbre*—and distinguishes them from mere noise, which lacks such dinergic harmony of different sounds. This dinergic harmony can be represented graphically, as in figure 15 where the vibrating string diagrams of figure 13 have been combined. (The small difference between the fifth's 2:3 = 0.666 ratio and the exact ratio of the golden section, 0.618:1 = 0.618, is 0.48, marked **d** in the drawing.) This kind of diagram, borrowed from the dinergic harmonies of vibrating strings, shall be used throughout this book to illustrate instances of similarly harmonious proportions.

Looking at the pattern of the keyboard in figure 16, we recognize its harmonious, golden proportions: there are 8 white keys, there are 5 black keys, and these black keys are in groups of 2's and 3's. The series 2:3:5:8 is, of course, the beginning of the Fibonacci series, the ratios of these numbers all gravitating toward the irrational and perfectly reciprocal 0.618 ratio of the golden section.

Our Western diatonic scales and chords are yet further instances of the 1:2, 2:3, and 3:4 proportions in the dinergy of musical harmonies. The two chief modes of Western scales, the minor (considered to be the sad one) and the major (associated with brightness) differ from each other only in the length of steps between certain of their intervals, just as the minor and major parts of the golden section differ from each other only in their lengths. And just as the joining of minor and major parts delights us in the visual harmonies of the golden section, so the joining of minor and major scales, called *modulations,* enchant us when we hear them in the movement of chords and melodies.

Every one of the minor and major scales has its own variants, or degrees—so-called dominants and subdominants—with their own sets of chords, and the relationship of these to their tonic counterparts again follows the above proportions. Dominants start and end on the fifth

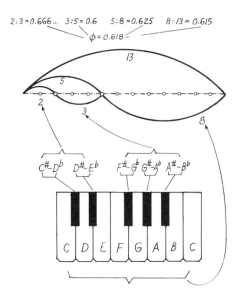

$2:3 = 0.666...$ $3:5 = 0.6$ $5:8 = 0.625$ $8:13 = 0.615$
$\phi = 0.618$

Fig. 16. Golden proportions of the keyboard.

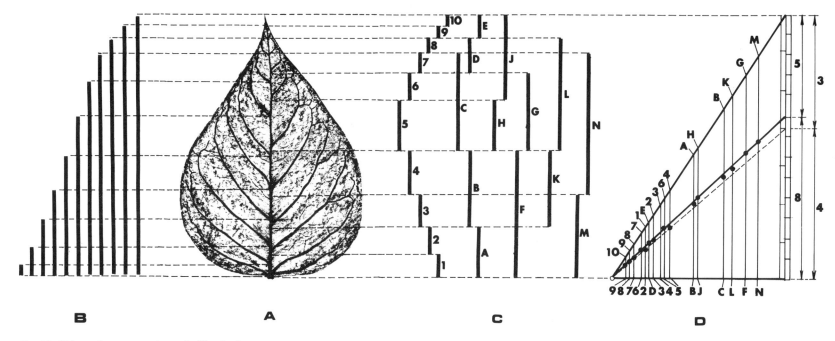

Fig. 17. Harmonious proportions of a lilac leaf.

interval from the keynote (the first note of the scale), and subdominants start and end on the fourth interval.

It would go beyond the limits of this presentation—and also beyond the limits of the author's expertise—to delve into further details of dinergy and analogies of visual proportions in musical harmonies. However, one more instance should be mentioned: counterpoint. In counterpoint, the dinergic joining of two or more different, often contrary musical lines unite and complement each other while preserving their own identities, much like dinergically joined spirals create visible harmonies in the daisy and the sunflower.[6]

Let us now look further into plant growth in the light of what we've seen about musical harmony. Figure 17 is a rubbing taken from the back side of a lilac leaf. Diagram **B** shows that the distances between the starting points of the veins along the midrib group themselves in a harmonic order, like organ pipes. The distances between consecutive veins marked 1 to 10 in diagram **C** form a similar series. These distances and their groupings have been lined up by neighboring pairs in diagram **D.** Their proportional relationships fall within the narrow limits of the golden section's proportional rate 0.618 . . . approximated by 5:8 = 0.625 (dash-dotted line) and the 3:4 = 0.75 proportion of the pentagonal Pythagorean 3-4-5 triangle, corresponding to the diapente and the diatessaron musical harmonies.

These relationships reveal growth patterns that are harmonious and dinergic in the sense that all the minors and majors (large and small veins and branches) unite with their neighbors in proportions limited to ratios of the same small whole numbers which create the root harmonies of music. Similarly dinergic and harmonious growth processes can be observed in the shaping of leaves other than the lilac.

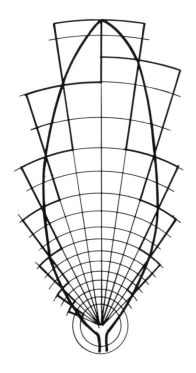

A Rhododendron

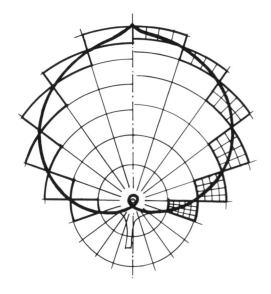

B Bloodleaf plant

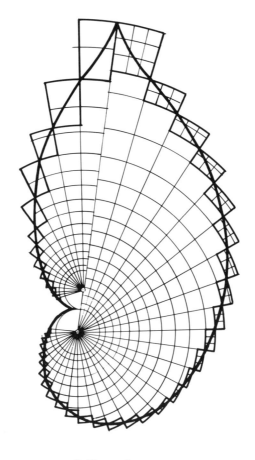

C Begonia

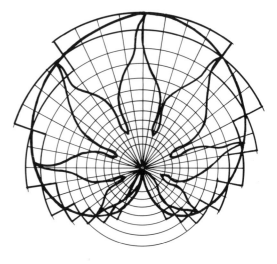

D Japanese maple

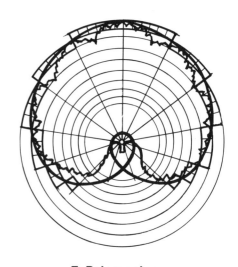

E Pelargonium

Fig. 18. Reconstruction of leaf outlines.

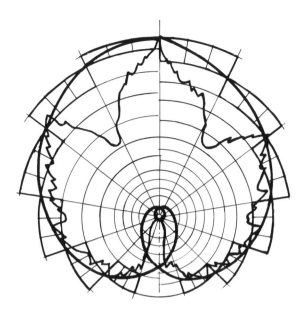

F Concord grape

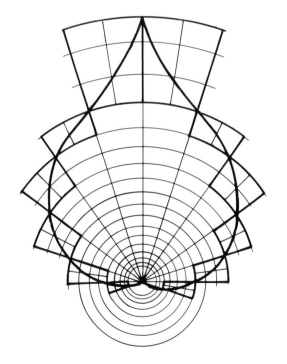

G Lilac

In figure 18 the outlines of a random selection of leaves were reconstructed with the dinergic method of combined radiating and rotating lines. If one looks at these patterns not as still pictures but as traces of the dinergic process which brought them into being, then the outlines of these leaves become their story, recorded in the silent language of patterns.

The outlines of the rhododendron leaf (**A**) start from the stem at the center with two rising circles, moving from one radius to the next. This pace increases to three circles at midspan, only to slow down again to two, winding up with approximately 1 1/3 at the tip of the leaf. This is the life story of the lanceolate leaf pattern.

The rounded or orbiculate leaf of the bloodleaf plant (**B**) is produced by a different pattern. The contour starts by crossing five circles while moving from the first radius to the second, and this rhythm gradually decreases to three, then to two circles, finally slowing down to one at maturation.

The two halves of the begonia leaf (**C**) develop at different rates, making it characteristically asymmetrical. Yet other varieties of harmonious patterns are created by diverse rates of development in the quinquefoliolate leaf of the Japanese maple (**D**), the lobed leaves of pelargonium (**E**), and the Concord grape leaf (**F**). In these latter two patterns, only the enveloping outlines have been reconstructed, though the component details could very likely be drawn in a similar manner. Finally, the cordate leaf of the lilac (**G**) starts by moving through four circles within the span of the first two radii, then it slows down to three, to two, and to one, only to take a last heroic spurt, rising three circles before the end.

These are just a few examples of leaf outlines, tracing pattern-forming processes similar to those that shape daisies, sunflowers, the flowers of edible fruits, and musical harmonies. They indicate that the same dinergic harmonies that delight our eyes in the shape of leaves and flowers also enchant our ears in the chords and melodies of music.

This chapter has ranged far and wide in the pursuit of dinergy, the energy-creating process that transforms discrepancies into harmonies by allowing differences to complement each other. Dinergy accomplishes this through the power of certain proportions, analogous to musical and root harmonies, well-known since antiquity, chief among them being the golden section.

The power of the golden section to create harmony arises from its unique capacity to unite the different parts of a whole so that each preserves its own identity, and yet blends into the greater pattern of a single whole. The golden section's ratio is an irrational, infinite number which can only be approximated, yet such approximations are possible even within the limits of small whole numbers. This recognition filled the ancient Pythagoreans with awe: they sensed in it the secret power of a cosmic order. It gave rise to their belief in the mystical power of numbers. It also led to their endeavors to realize the harmonies of such proportions in the patterns of daily life, thereby elevating life to an art.

Fig. 19. Dinergic construction of spider web joining radiating, straight lines with rotating spiral ones.

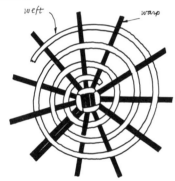

Fig. 20. Dinergy of basket weaving. Firm straight strands of warp support flexible spiral strands of weft.

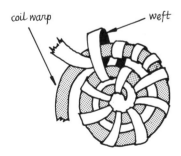

Fig. 21. Coil binding, in which both weft and warp are flexible. The weft binding is sewn along radiating lines around the warp, which takes the form of a spiral coil.

CHAPTER 2: Dinergy in the Crafts

Orb-weaving spiders construct their webs by starting with straight threads that come together in a center. Then they spin spiral strands around the straight ones, rotating in ever wider orbits. (fig. 19)

Basket weavers work in a similarly dinergic pattern. First, a number of firm strands—the warp—are bound together at one point which shall become the center of the basket. (fig. 20) Then flexible strands—the weft—are passed over and under the warp strands in a rotating fashion. In coil-binding, a firm but flexible coil takes the place of the straight warp, and it is sewn together along radiant lines with thin weft material, threaded through a needle. (fig. 21) Because of the dinergic nature of the work process, one can easily reconstruct basket outlines the same way we reconstructed leaf outlines and the spirals of the sunflower and daisy.

Two basketry hats (fig. 22) woven by American Indian people of the Pacific Northwest have been reconstructed in figure 23. The shaded triangles show successive stages of the dinergic work process, as they indicated successive stages of growth in the plant. The centers of the two spirals that make up the curved outline are not located at random. In shape **A** these centers coincide with the center of the hat's crown. In shape **B** they are at golden section points within two squares *over* the hat, identical with the two squares that enclose the hat itself. (See golden section constructions and wave diagrams.)

One might be tempted to attribute such hidden order to chance, but the frequency of such chance order is extraordinary. Fourteen samples of such hats were measured in the Thomas Burke Memorial Museum of the University of Washington, eight of the concave type A and six of the convex type B, and all show, in varying degrees and in different ways, close approximations to the proportions of the golden section and to the Pythagorean triangle's proportions, corresponding to the musical harmonies of fifth-diapente and fourth-diatessaron, respectively.

The proportional pattern of the concave hats (type A) woven out of cedar bark and beargrass by Makah Indian women, and common among other Nootka people, shows further variations of golden dinergy. Figure 24 demonstrates four of these relationships with constructions of the

Fig. 22. **A.** Concave hat woven of cedar bark and beargrass by the Makah and other Nootka peoples.

B. Convex hat woven from spruce roots, popular with Tlingit, Haida and Kwakiutl peoples.

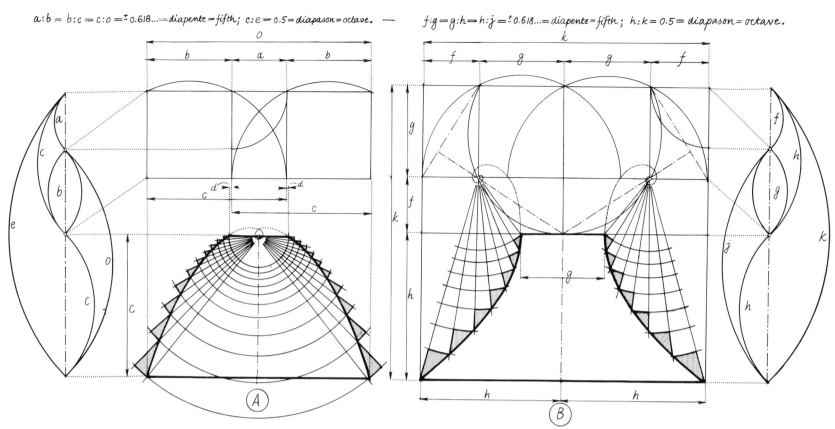

Fig. 23 Basketry hat shapes reconstructed by dinergy of radiational and rotational work processes.

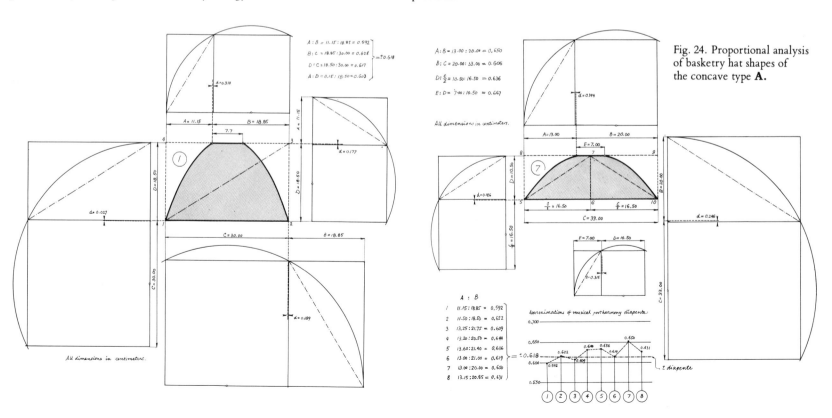

Fig. 24. Proportional analysis of basketry hat shapes of the concave type **A.**

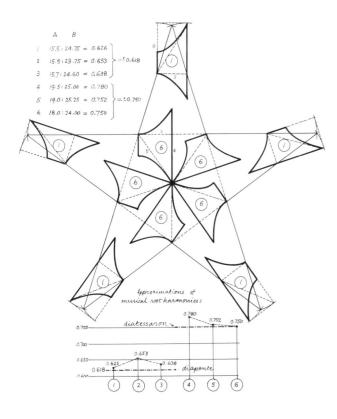

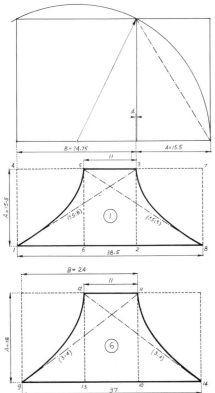

golden section around one of the tallest and one of the lowest of these hats. The overall shape of the taller hat fits neatly into a single golden rectangle (1-2-3-4) while the short hat approximates two such rectangles (5-6-7-8 and 6-7-9-10).

For each of the eight type-A hats measured, the proportional relation of neighboring parts **A** and **B** was tabulated (see chart at bottom of fig. 24). The closeness of these proportions to those of the golden section is again demonstrated by classical golden section constructions around the shapes. The differences between actual and theoretically exact logarithmic ratios are so small they had to be exaggerated to be visible.

Figure 25 shows two typical examples of the convex type-B hat, woven out of spruce roots and popular with Tlingit, Haida, and Kwakiutl people. Three of each type were measured. The shallow hats tend to approximate golden proportions, and the tall ones tend to show the **3:4** proportions of the Pythagorean triangle. This can be seen from the dash-dotted lines connecting opposite ends of the top and bottom diameters. In the shallow types, these lines coincide with the diagonals of a golden rectangle, while in the tall ones they are identical with the longest side (hypotenuse) of the Pythagorean triangle. The accuracy of these proportions can be appreciated from the tabulation and diagram, as well as from the construction of the golden section above the shallow hat.

To make the harmonies of these patterns more visible, these two shapes are also shown in their correspondence to the pentagonal star, the taller shape fitting neatly into the Pythagorean triangles within the central pentagon, and the shallow shapes aligning with the triangles at the tips of the pentagram.

The lines of these hats suggest the sinuous energy of the drawn bow as well as the graceful curves of flower and leaf outlines. These harmonious and dinergic proportions emerge as naturally from the hands of these preliterate basket weavers as bows bend and plants grow. And basket weaving is not the only craft for which this holds true.

Warp and woof

The weaving of fabrics is as dinergic a process as the weaving of baskets, with the difference that in fabric, both warp and woof move along straight lines, crossing each other, the threads of the warp being held evenly within a frame, while the threads of the woof are laced over and under them. (See diagram in fig. 26)

Weavers belonging to different cultures show the preference for simple harmonious proportions that basketmakers do. An East Prussian carpet reveals the proportions of a golden rectangle in its overall shape (fig. 26) as does the Mexican woven pattern in its repeated diamond shapes. (fig. 27)

Fig. 25. Proportional analysis of basketry hats of the convex type **B.**

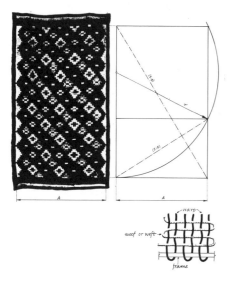

Fig. 26. Peasant carpet from Eastern Prussia.

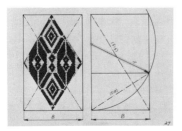

Fig. 27. Woven Mexican pattern.

The masterfully woven ceremonial blankets of Northwest American Indian people show the same preference in overall shape, as well as in the articulation of the design, carried through consistently even to minute details. Figure 28 shows golden section constructions drawn around a Chilkat blanket. These proportions relate the short sides to half the width, as well as to the middle height, the latter two dimensions usually being equal in these blankets. The minute difference between exact ϕ ratios and the actual ratios in this blanket are again marked with **d.** The blanket's short sides allow it to be wrapped conveniently around the body, but the dimensions of the sides are never arbitrary. The seventeen samples examined all approximate these same proportions.

The pictorial pattern of these blankets often articulates as a wider central panel flanked by two narrower, lateral ones, as in this sample. The golden section construction above the blanket shows that the width of these unequal neighboring elements—**D** and **E**—are also in golden relationship, containing the overall shape within two reciprocal golden rectangles (1-2-3-4 and 3-4-5-6), jointly producing the familiar $\sqrt{5}$ rectangle.

Fig. 28. Woven Chilkat blanket.

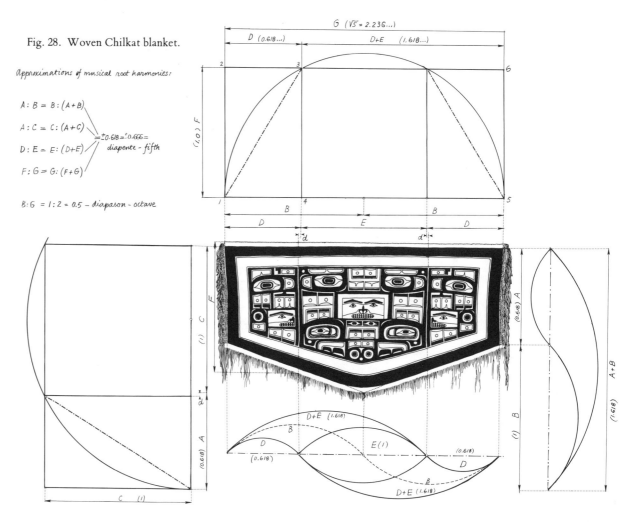

Approximations of musical root harmonies:

$A : B = B : (A+B)$

$A : C = C : (A+C)$

$D : E = E : (D+E)$ $= \pm 0.618 = 0.666 =$ diapente - fifth

$F : G = G : (F+G)$

$B : G = 1 : 2 = 0.5 -$ diapason - octave

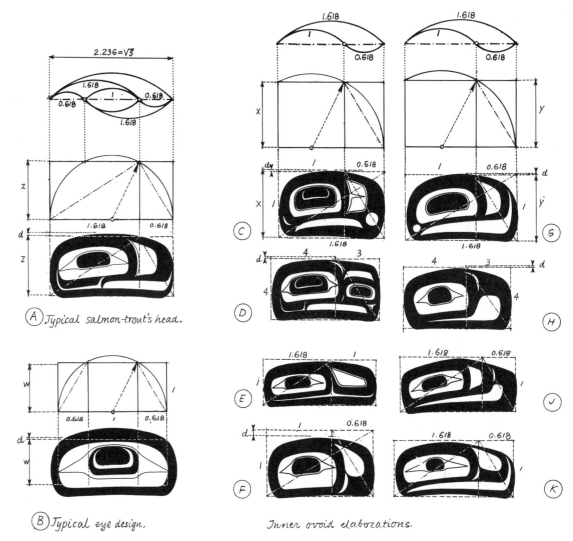

Fig. 29. Proportions of ovoids in Northwest Indian art of U.S.

Similar proportions extend to details, such as the eyes and eye-like ovoids, which are characteristic features of Northwest American Indian art. Exceptions to these proportions occur when designs are adjusted to fill the space available.[7]

In figure 29, a "salmontrout's head" design (**A**) is shown to be contained within a $\sqrt{5}$ rectangle (corresponding to the musical harmony of octave or diapason) and consisting of two reciprocal golden rectangles, the larger one of which contains the eye. The typical eye design (**B**) also shares the same relationships both in the shape of the inner ovoid, which represents the white of the eye, and in the size of the iris. Variations of inner ovoid elaborations also reveal a preference for the golden relations of neighbors, as in details **C, E, F, G, J,** and **K.** Two variations—**D** and **H**—combine a square with two Pythagorean 3:4 triangles.

Hands and wheels

A potter's hands press the clay toward the center of the wheel upon which the clay is placed, while the wheel is turning around, thus providing the dinergy that shapes the vessel. Figure 30

Fig. 30. Lung Chuan celadon vase
(Sung dynasty, 960-1279).

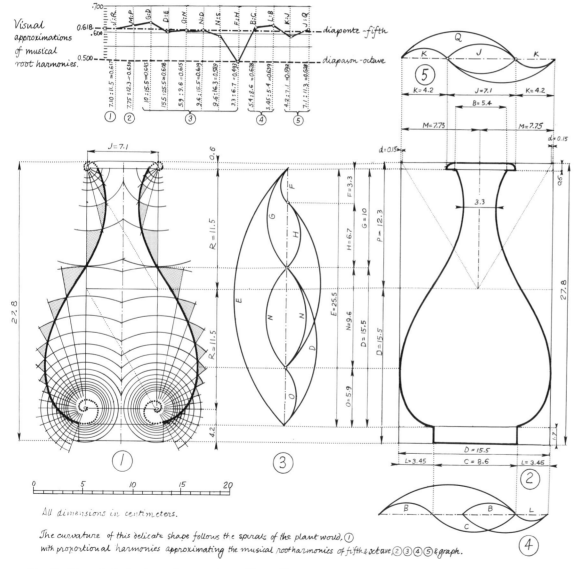

Fig. 31. Dinergic derivation and proportions of Lung Chuan vase.

shows a Chinese Lung Chuan celadon vase from the time of the Sung dynasty (960–1279 A.D.). In figure 31, the outlines of this graceful shape have been derived with the dinergic method used earlier for the reconstruction of plant shapes. This derivation shows that the centers of the four logarithmic spirals that make up the outlines are located at the corners of two rectangles closely approximating golden proportions, (7.1:11.5 = 0.617± = 0.618...). Even without knowing these numerical relationships, the similarity of these contours to leaf outlines would suggest the relatedness of this shape to patterns of organic growth. This suggestion is verified by the wave diagrams and the graph, which record how the proportional relationships between the main features approximate musical root harmonies.

The entire shape is contained in a rectangle consisting of a square—which encompasses bulk and base—and two golden rectangles, the heights of which correspond to the height of the slender neck. These proportional harmonies are shared also between the height of the upper part of the vase (**G**) and the height and girth of the bulky body (**D**); between the latter and the height (**E**); and between the diameters of brim (**B**) and base (**C**). (See diagrams 1, 2, 3.) The contrast between slender neck and broad base, between narrow top and wide girth, could hardly be greater. Yet these contrasting elements are united in graceful harmony, by sharing the dinergic relations of neighbors.

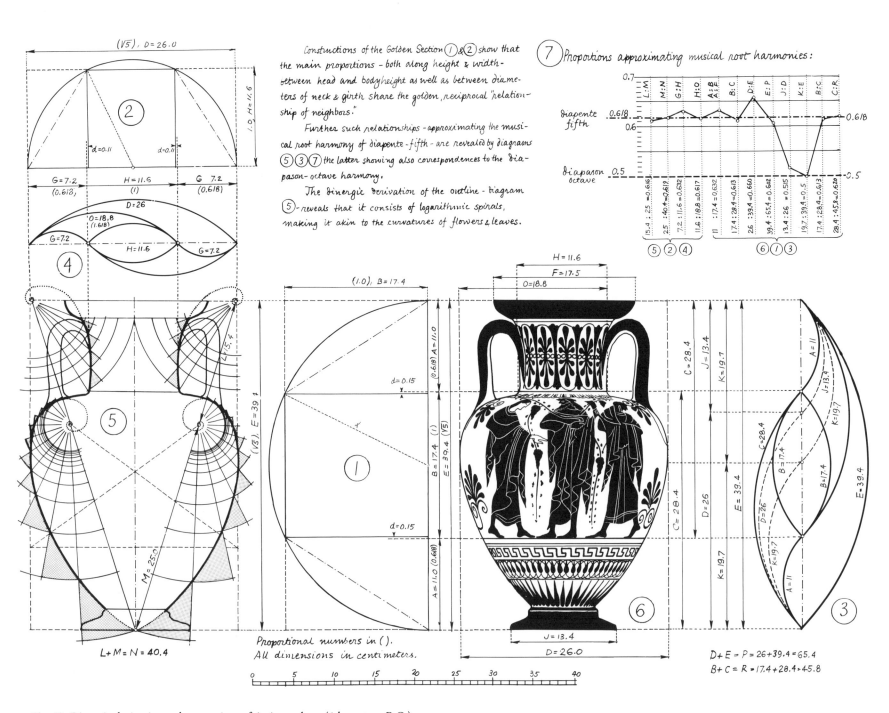

Constructions of the Golden Section ① & ② show that the main proportions - both along height & width - between head and body height as well as between diameters of neck & girth share the golden, reciprocal "relationship of neighbors."

Further such relationships - approximating the musical root harmony of diapente - fifth - are revealed by diagrams ⑤ ③ ⑦ the latter showing also correspondences to the diapason - octave harmony.

The Dinergic Derivation of the outline - diagram ⑤ - reveals that it consists of logarithmic spirals, making it akin to the curvatures of flowers & leaves.

⑦ Proportions approximating musical root harmonies:

Proportional numbers in ( ).
All dimensions in centimeters.

$D + E = P = 26 + 39.4 = 65.4$
$B + C = R = 17.4 + 28.4 = 45.8$

Fig. 32. Dinergic derivation and proportions of Attic amphora (6th century B.C.).

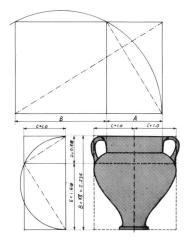

Fig. 33. Cretan krater.

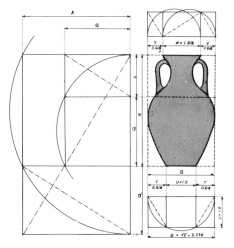

Fig. 34. Cretan amphora.

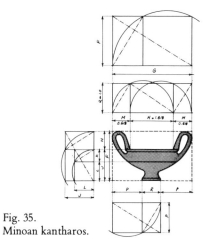

Fig. 35.
Minoan kantharos.

If we move from the China of Sung times to Greece in the sixth century B.C., we find the same dinergic work processes creating dinergic proportional relationships in the masterpieces of classical Greek pottery. An Attic amphora depicting the legend of Hercules and Pholus is outlined in figure 32. The dinergic derivations of the shape show that two centers of the logarithmic spirals are in the upper corners of the golden rectangle which contains the whole shape, while the lower spirals' centers are at golden section points along the diagonals within the two halves of this rectangle. Furthermore, these two lower centers are at the very tips of a pentagonal star which touches the two sides and the base of the enveloping golden rectangle.

The golden section's constructions show how closely the main neighboring elements of this design approximate reciprocally shared golden proportions. In diagram **1,** this relationship is shown to exist between height of head (**A**), height of central picture-band (**B**), and total height of the vessel (**E**). Diagram **2** and wave diagram **4** show the same relationship shared between shoulder width (**G**) and diameters of neck (**H**) and girth (**D**). (Differences between actual and theoretical ϕ ratios are indicated with **d**.)

Wave diagram **3** illustrates yet further harmonious proportions in this amphora. Golden relations approximating the musical harmony of fifth-diapente exist also between diameter of brim (**F**), height of head (**A**), and height of bulk (**C**). The octave-diapason relation shows up in the location of girth (**K**) at half the total height (**E**); it is also approximately by the diameter of the base (**J**) being half the diameter of the girth (**D**).

Such consistency of proportional relations in classical Greek pottery was demonstrated in great detail by the American scholar Jay Hambidge in the early 1920s.[8] This book is very much indebted to Hambidge's pioneering work. Since the 1920s, however, much archeological and ethnographic research has shown that such proportions were not restricted to the classical Greeks and their Egyptian teachers, as Hambidge thought, believing their advanced arithmetic and geometrical knowledge to be the source of these proportions. To our earlier examples from other cultures, let us now add some pottery unearthed from graves and other excavations dating back to the Cretan and Mycenaean cultures of the late Bronze age, which preceded the flourishing of classical Greek arithmetic and geometry by about a thousand years.

Figure 33 is a Cretan krater which is contained in two sets of reciprocal golden rectangles, totaling a length of $\sqrt{5}$, as if these rectangles had been jointly rotated around the longitudinal axis of the vessel. The smaller of the golden rectangles contains shoulders, head, and handles, while the larger ones correspond to the bulk of the body. In addition, the width of the opening (**A**) and the total height (**B**) share the same golden relation of neighbors.

The Cretan amphora in figure 34, unearthed from a grave, is contained in a golden rectangle (**Q × R**), within which shoulders, head, and handles fit into reciprocal golden rectangle **Q × S**. The proportions of two reciprocal golden rectangles exist in the relation of brim (**W**) to shoulders (**V**) to girth (**Q**); it also exists in the relation of base (**U**) to its recesses (**T**) to girth (**G**).

Figure 35 shows a Minoan kantharos contained in a single golden rectangle, **F × G**. The width of the handles (**M**) relates to the distance between them (**N**) as the sides of two reciprocal golden rectangles do: 0.618 and 1.618. Golden dinergic relations are also shared between the width of base (**R**) and its distance from edge (**P**); between the height of handles (**H**) and the height of body (**J**); between body height (**J**) and the total height including handles (**F**); between the upper and lower parts of the body (**K** and **L**); and between the lower part's height (**L**) and the height of the entire vessel (**F**).

Fig. 36. Earthenware pot from Acoma Pueblo.

Figure 36 is an exquisite earthenware pot from Acoma Pueblo. The shape is a popular one with Zuni and Pueblo people and is made in different sizes and proportions. The outlines of three of these—number **3** being the vessel pictured—are shown superimposed over each other in figure 37. Dotted lines indicate the two golden rectangles (of various sizes) into which all of these shapes tend to fit, as if such rectangles had been rotated along the axis of each vessel, as in the Cretan krater in figure 33. This supports the belief of psychologists John Benjafield and Catherine Davis that "the Golden Section is an important regulating principle at least in some forms of folk art."[9]

Tabulation and graph demonstrate that here too, as in the other vessels we examined, harmony is created by a conscious or unconscious tendency to unite the diverse elements of the shape, through shared proportions, ranging from the 0.5 and 0.75 rates and gravitating—scissor fashion—around 0.618, these three proportions corresponding to basic musical root harmonies.

In figure 38 the Acoma pot number **3** has been rendered in more detail by dinergic derivation of its outline (left). The centers of the logarithmic spirals constituting the outline fall along the diagonals of the golden rectangles which contain this outline. The triangular diagram, which includes the rhythmic wave diagrams of some of the proportions investigated, reveals at a glance the harmony created by the unity of minor and major parts, achieved by their shared proportional limitations.

These curved diagrams are like graphic images of a reciprocal echo in which the whole reverberates the call of every part and every part responds to the call of the whole. In another sense, this diagram expresses the dance-like rhythm of the whole design, which is set swinging by the harmonious dinergy of the work process as well as by the golden dinergy of its proportions.

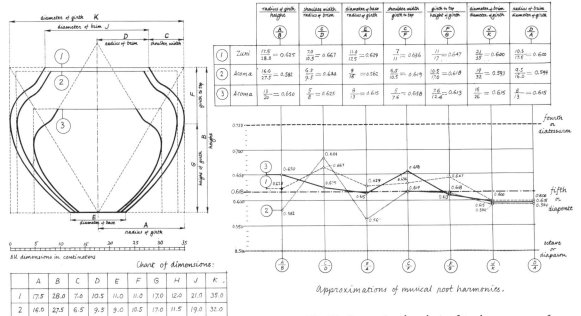

Fig. 37. Proportional analysis of earthenware pots from Zuni and Acoma Pueblos.

This exquisite vessel is created by dinergic pattern-forming processes unfolding upon three different levels. The first level is the dinergic work process itself, demonstrated below as a union of rotating and radiating work components, generating the spiral outline. The second level of dinergy is in the golden proportional "relations of neighbors" lined up in the triangular diagram at the right. The arcs have been added to emphasize the rhythmic harmony which unites all small and large elements of the design in dance-like dinergy. Two near-perfect "relations of neighbors" are shown above and below the vessel. The third level of dinergy is the painted pattern; a remarkable combination of rectangular spirals, moving in opposite directions, in and out, up and down, right and left.

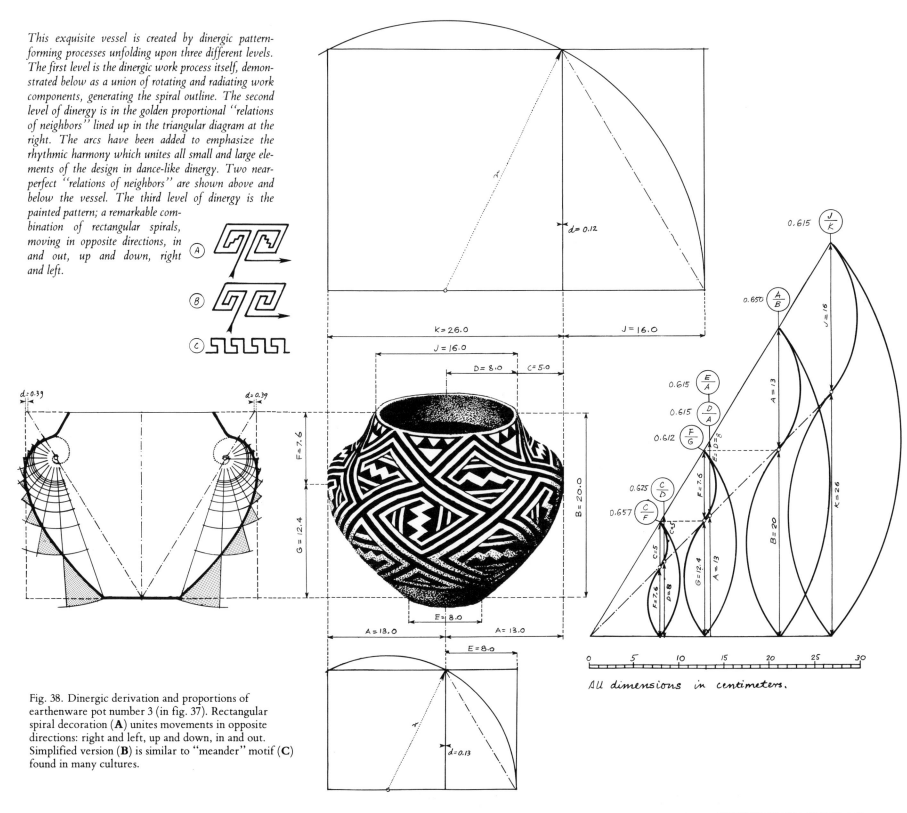

Fig. 38. Dinergic derivation and proportions of earthenware pot number 3 (in fig. 37). Rectangular spiral decoration (**A**) unites movements in opposite directions: right and left, up and down, in and out. Simplified version (**B**) is similar to "meander" motif (**C**) found in many cultures.

When one looks closer at the fascinating painted pattern of this vessel, one becomes aware of yet another level of dinergic pattern formation. The intricately woven zig-zag lines, which all move in different directions—in and out, up and down, right and left—are all parts of one single line, like vines growing from one stem. One element of this pattern is shown at the left in simplified detail (**A** and **B**). Similar meandering patterns symbolizing dinergic unity occur in the art of many cultures, in countless varieties, a simple version of which (**C**) can be seen at the base of the Attic amphora, figure 32.

This threefold dinergy of painted pattern, proportions, and work process conveys a strong sense of oneness between the Indians and all that exists around them, expressed by the Oglala Sioux sage Black Elk in the following way: "We know that we are related and one with all things of the heaven and the earth . . . the morningstar and the dawn which comes with it, the moon of the night and the stars of the heavens. . . . Only the ignorant person . . . sees many where there is really one."[10]

This basic belief throws light upon the dignity which a work of pottery such as this one radiates. In the eyes of the people who make and use such earthenware pots, these are not merely dispensable utensils. Rather, each of these pots is felt to have a life and a spirit of its own, similar to its maker's, and consequently these pots are made and used with the respect accorded to a living being. Anthropological research since the end of the last century refers to this belief as *animism.* According to anthropologist F. H. Cushing: "The noise made by a pot when struck or when simmering on the fire is supposed to be the voice of an associated being.... That [the being departs when the pot is broken] is argued from the fact that vessel when cracked or fragmentary never resounds as it did when whole."[11]

We may smile at such beliefs, thinking them to be antiquated, naive, and unscientific. Nevertheless, there is an element of truth here. There is a wholeness in which everybody and everything is related, just as the diversities of these patterns are related. All people and all things are indeed neighbors.

CHAPTER 3: Dinergy in the Arts of Living

Fig. 39. Polynesian spiral tattoo, New Zealand.

Fig. 40. Hopi mother-earth symbol.

Fig. 41. Coin from Knossos, Crete.

Tangible and intangible patterns

The dinergic nature of symbols, bridging the gap between tangible and intangible patterns, was briefly touched upon earlier. Here it may be pointed out that the word *symbol* reveals its own dinergic origin, since it comes from the union of two Greek words: *sun,* "together," and *ballein,* "to throw," like *ball* which also comes from the latter root.

The first symbol we examined was the pentagram, a tangible pattern that helped the Pythagoreans grasp the intangible realities of harmony and health. The fact that the pentagram is still used as a symbol of good portent shows that the dinergy of certain symbols is both timeless and universal. C. G. Jung called such symbols *archetypal,* and he devoted a lifetime to the study of their central importance in shaping patterns of human behavior.

Because harmonious patterns created out of tangible materials are readily completed, where-as the intangible patterns of the arts of living are constantly in the making, one could perhaps look upon the harmonious patterns of the arts and crafts as symbols, metaphors and models of similarly harmonious behavior patterns yet to be created in the arts of living.

It was suggested earlier how the interrelatednes of diverse elements in a Pueblo pot expresses the Indians' unity with nature. The same unity is expressed by the Maoris of Polynesia in their concepts of *mana* and *tapu,* the latter being a variant of *taboo.* The noted Danish anthropologist Kaj Birket-Smith describes the experience of mana as a strong feeling that "life is unity, in which not only gods, but also things, which to us are lifeless have a part." Mana is thus a direct experience of "the sacred force that permeates existence." *Tapu* is the Maoris' word for their sacred responsibility to comply with mana; a supreme law.[12]

Mana and tapu are expressions of the Maoris' sense of relatedness and oneness with the universe around them, and just as the American Indians express this feeling with spiral patterns, so do the Maoris. Everywhere in Maori art spirals abound. They are carved into wood and stone, painted, and even tattooed on the body in the hope that the mana power attributed to these symbols will save their wearers from trouble and untimely death. (fig. 39)

Among American Indians, it is not only the Pueblo people who use spiral patterns. They can also be found engraved into the rocks in some of the oldest continuous human habitations in America, the villages of Oraibi and Shipaluovi, as well as in the ruins of Casa Grande, near Florence, Arizona. (fig. 40) The Hopi Indians refer to these patterns as Mother Earth symbols, as *Tapu'at* (mother and child), or as symbols of emergence and rebirth. Anthropologists report that similar symbolic meaning is attached to the same symbols by many other Indian people all over the Americas.[13]

Identical spiral patterns have been discovered in many other parts of the world, dating from prehistoric times. On the Mediterranean island of Crete, a 3000-year-old coin (fig. 41) shows exactly the same spiral as the American Tapu'at, but in Crete this pattern represented the famous palace of the Minoan kings, the Labyrinth. Legend says that the Labyrinth was once the lair of the Minotaur, a mythical beast, half bull and half man, which was a symbol of fertility.

Fig. 42. Cretan mother goddess, or her priestess.

Fig. 43. Hermes-Mercurius.

Fig. 44. Rain ceremony or "Tree of Life" after rock painting by Wahungwe tribal artist, Zimbabwe.

An apparently universal symbol of fertility is the snake, whose coiled body may very well have contributed to the creation of archaic spiral patterns. The snake lends itself to this symbolic role also because it strikes erect—a reminder of phallic power. The Great Mother goddess of Crete and her priestesses are frequently represented holding snakes.(fig. 42)

Hermes or Mercury, messenger of the Greek and Roman gods, had two snakes winding around his magic wand, the caduceus, tool of his healing power, and even today professions connected with healing use this dinergic symbol as their emblem. (fig. 43) Hermes was also the guide of the departed to the afterworld, and the intertwined spiraling snakes of his magic wand may also symbolize the intertwined mystery of life and death.

The intangible dinergy of life and death—interwoven with fertility—also appears in the tribal art of the Wahungwe people of Zimbabwe. Figure 44 is a drawing made from a tribal artist's painting, showing a tree rising from the dead body of a woman into the sky, where a goddess and a giant snake cause fertilizing rain to gush forth out of the roots of the tree. In the biblical creation story, woman, tree, serpent, and fertility are likewise intertwined, but here a further intangible dinergy enters: the knowledge of good and evil.

Fig. 46. Prehistoric carved stone ball, believed to have been used for divination.

Fig. 47. Spirals or whorls in fingerprints.

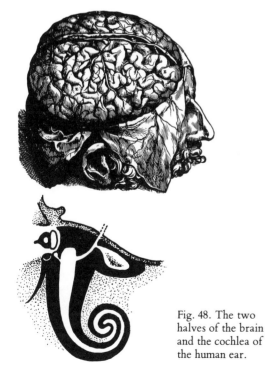

Fig. 48. The two halves of the brain and the cochlea of the human ear.

Fig. 45. Prehistoric spiral mazes at New Grange, Ireland. Threshold stone, *left;* bas relief carved into wall, *right.*

Intertwined spiral mazes from Neolithic times, identical with the Cretan Labyrinth, the Maori tattoo and the American Indian Tapu'at, are carved into the rocks of barrow tombs in New Grange, Ireland. (fig. 45) These double spirals have been interpreted as symbols of death and rebirth, because as one follows the line coiling inward, one finds another line coming out in the opposite direction, suggesting both burial in the tomb and emergence from the womb: the dinergy of life and death.[14]

The same double spiral design shows up with remarkable clarity carved into stone balls which have been unearthed near prehistoric passage graves at Glas Towie, Scotland. (fig. 46) As anthropologist Loren Eisley has written, "Neanderthal man had, we now know after long digging, his own small dreams and kindnesses. He had buried his dead with offerings—there were evidences that they had been laid, in some instances, upon beds of wild flowers."[15]

The double spirals of prehistoric graves are witness to the kindness as well as to the creative concerns of our prehistoric ancestors. These patterns reveal the energy-creating power of symbols suggested by the word dinergy. The tangible unity of the dinergic spiral lines brings home the realization that death and life are in some mysterious and unfathomable way interlocked. The sense of awe that follows this realization sparks new energy from the embers of grief.

Our dinergic endowment

One can find dinergic spirals much closer at hand than in prehistoric tombs—for instance, in the minute whorls of our fingertips. (fig. 47) Dinergy is also evident in many other patterns of our physical and mental constitution.

We have two eyes and we see two images, which are united in the brain into a single, three-dimensional stereoscopic vision. We have two ears, receiving signals from two opposite directions, which are transmitted through the spiral-shaped cochlea of the inner ear to unite in the brain as stereophonic sound. (fig. 48)

The entire nervous system is a dinergic double structure, consisting of peripheral and central component systems, again united by the brain. The brain itself consists of two hemispheres, their integration being performed by centrally located organs within the brain. The bold and extraordinary theory of Julian Jaynes traces the origin as well as the present-day problems of consciousness to a breakdown in the double structure of our minds.[16]

Is it pure coincidence that on the molecular level the joint three-dimensional spiral pattern of the double helix—matching the double snakes of Hermes' magic wand—was found a few years

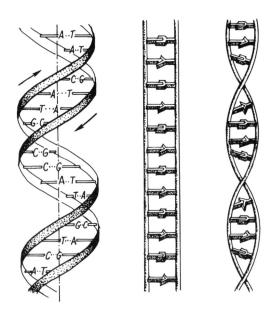

Fig. 49. Models of DNA double helix.

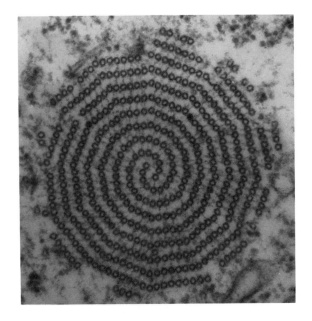

Fig. 50. Axoneme, the core of a single
axopod, shown in cross-section. Enlarged 90,000 diameters.

ago to be the true shape of the DNA molecule, which contains within its miniature coded pattern the master plan of the entire future development of living organisms? (fig. 49) It was discovered even more recently that some of the most minute and important elements within living cell structures (such as red and white blood corpuscles) group themselves in double spiral patterns. The cores of these microtubules, referred to as *axonemes* and seen here enlarged in the electron micrograph, figure 50, are a faithful match to the double spirals of prehistoric tombs, the tattoos of the Maoris, and the Mother Earth patterns of the American Indians.

Whatever else there may be behind such "coincidences," it is hard to avoid the conclusion that we are looking at one of nature's most basic pattern-forming processes, here referred to as dinergy. Seeing the hidden and harmonious order built into body and mind, as it is built into every flower and leaf, mirrored by the crafts, and echoed by music, one wonders at the origin of the disharmony and disorder that mars our civilization.

Indeed, our fascination with so-called "primitive" cultures appears to spring from our longing for the lost dinergic relatedness which was once ours, when we ourselves were still "primitives." Of course, we very much need science and technology, but we do not need the fragmentation and separation that have come with the differentiations of our civilization. Perhaps the disharmonies and disorders are with us not because our culture has grown up, but because we have not yet grown up. Western civilization is still in its adolescence. Our violences and worries may be but growing pains.

We do not have to go back to preliterate tribal cultures to find true dinergic relatedness to nature and the universe. St. Francis of Assisi, in one of his famous canticles, addresses the Sun, Air, Fire, Wind, and Water as his brothers; he sings about Moon and Stars as his sisters and he praises the Earth as his mother.[17]

A sense of our dinergic relatedness to the mysteries of this universe and our participation in its mana has inspired many scientists. Albert Einstein wrote: "Enough for me the mystery of the eternity of life and the inkling of the marvelous structure of reality, together with the single-hearted endeavor to comprehend a portion, be it ever so tiny, of the reason that manifests itself in nature."[18]

The psychologist William James saw the essence of true religion as "the belief that there is an unseen order, and that our supreme good lies in harmoniously adjusting thereto."[19] Another psychologist, Abraham H. Maslow, refers to the "peak experience" as "a clear perception that the universe is all of a piece and that . . . one is part of it, one belongs in it." From this follows a sense that "the sacred is in the ordinary . . . that it is to be found in one's daily life, in one's neighbors, friends and family, in one's backyard. . . . To be looking elsewhere for miracles is to me a sure sign of ignorance that *everything* is miraculous."[20]

CHAPTER 4: Timeless Patterns of Sharing

Basic arts of sharing

Sharing is not only a basic pattern-forming process and an art; it is also a condition of life. With every breath of air, with every sip of water or bite of nourishment, we share the resources of the earth. In *The World As I See It,* Albert Einstein says: "A hundred times every day I remind myself that my inner and outer life depend on the labours of other men, living and dead, and that I must exert myself in order to give in the same measure as I have received and am still receiving."[21] This is the reciprocal sharing of the Golden Rule and of the golden section.

Before we share in the labors of others we share communications with them. Speech is one of our oldest means of communication, a basic art through which we share our feelings, our needs, and our thoughts with others. The development of speech goes back perhaps a million years, so we don't know much about its beginnings. But the discovery that many languages have a common origin (such as the Indo-European root language, now lost, after having given birth to many modern languages) gives us a glimpse of the beginnings of the art of sharing through speech. The word *mother,* which is similar in many languages, has been traced back to the *ma* sound uttered by all babies.

Even the art of thinking originates in sharing through talk. As Don Miguel de Unamuno says: "To think is to talk to oneself and each one of us talks with himself thanks to our having had to talk with one another. Thought is inward language and the inward language originates in the outward. Hence it results that reason is social and common."[22]

We share communication also without words, using the silent speech of body language. We all know that a friendly smile, a frown, a shaking of the head or a handshake at times tells more than many words can.

The philosopher Susan Langer speaks movingly about the effect of sharing the experience of touch and discovering thereby the shared meaning of words, quoting a famous passage from Helen Keller's autobiography, where she describes how her teacher took her to a well. "Someone was drawing water and my teacher placed my hand under the spout. As the cool stream gushed over my hand she spelled into the other the word *water*, first slowly, than rapidly. I stood still, my whole attention fixed upon the motion of her fingers. Suddenly I felt a misty consciousness as of something forgotten . . . and somehow the mystery of language was revealed to me. I knew then that w-a-t-e-r meant the wonderful, cool something that was flowing over my hand. That living word awakened my soul, gave it light, hope, joy, set it free!"[23]

Both teaching and learning are essentially experiences of sharing. Good teachers have the magic capacity to share themselves and their devotion to their subject with their students. "The greatness of teachers is not measured by how much they know, but by how much they share," said the Reverend Jesse Jackson at a conference dealing with violence in our cities.[24]

Children learn what they see around them. If they see mostly egoism, they learn to be egoists. If they see sharing, they learn how to share, and—since learning is sharing—they learn to read, write and count in the bargain.

Shared human enterprises have always found expression in shared body movements, as figure 51, a drawing after a prehistoric painting, illustrates. Here early hunters or warriors are seen

Fig. 51. Marching hunters or warriors. After prehistoric cave painting at Casulla Gorge, Spanish Levant.

Fig. 52. Ring dance associated with cult of Hercules, inside Greek terra cotta cup (6th century B.C.).

Fig. 53. Dancing apsaras (heavenly nymphs) from Indian temple frieze (12th-13th centuries A.D.).

raising their bows and arrows as they stride in dance-like gestures toward their shared enterprise. Even thousands of years later, just by looking at them one can feel the mana-like energy emanating from their shared movements.

In a Greek cup painting from the sixth century B.C., we see dancers all enthusiastically dancing the same dance as they circle around the hero Hercules, who struggles with a sea monster. (fig. 52) *Enthusiasm* is a word that means literally "god within," (in Greek, *en,* "in" and *theos,* "god"). Enthusiasm is sharing in an energy that is conceived of as divine, as mana. We find movements expressing similarly shared religious emotions in countless other sacred dances, such as the dance of apsaras (semi-divine beings) in an Indian temple carving. (fig. 53)

We all become enthusiastic when we dance. With or without belief in divine beings, the mana of shared rhythms carries us in the waves of the dance. (fig. 54)

Fig. 54. Bulgarian folk dancers.

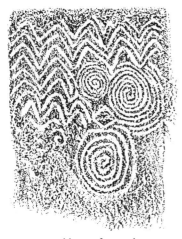

Fig. 55. Rubbing of carved design from passage grave in New Grange, Ireland.

Fig. 56. Geometric patterns made by Brazilian Indians. Top design represents bats, bottom design respresents bees.

One kind of body language leaves visible traces: writing. The art of writing is a form of communication that developed from patterns to which shared meanings were attributed. For instance, zig-zag lines found side by side with spiral patterns engraved in the walls of the prehistoric passage graves of New Grange, Ireland are looked upon by scholars as the first representations of water, believed to signify the earliest form of the letter M.[25] (figs. 55, 61)

In tribal cultures where writing has not yet developed, simple geometric patterns are used to share ideas. The anthropologist Franz Boas writes: "It is remarkable that in the art of many tribes the world over, ornament that appears to us purely formal is associated with meanings.... Geometrical patterns of the Brazilian Indians represent fish, bats, bees and other animals, although the triangles and diamonds of which they consist bear no apparent relation to these animal forms.[26] (fig. 56)

Even highly complex geographic information is at times passed along within tribal cultures by such pattern-images. The patterns on a rawhide bag made by North American Arapaho Indians (fig. 57) are said to represent a village surrounded by mountains (double frame), dotted with lakes (squares in frame), tents (dark triangles) and a buffalo path running right through the center (dotted line).

Sometimes long stories are shared by members of a tribe, with the help of a single pattern. For instance, the painted Zuni bowl in figure 58 narrates the story of "Cloud Alone": "When a person does not go to the dances, when they dance for rain, after her death she goes to the Sacred Lake and when all the spirits of the other dead people come back to Zuni to make rain, she cannot go but must wait alone, like a single little cloud left in the sky after the storm clouds have blown over. She just sits and waits all alone, always looking and looking in all directions, waiting for somebody to come. That is why we put eyes looking out in all directions."[27]

Fig. 57. Arapaho Indian rawhide bag. Pattern gives geographic information about village.

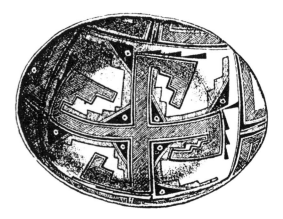

Fig. 58. Zuni bowl. Decoration tells story of "Cloud Alone."

When in the course of historical development large urban population centers arose, as in Mesopotamia, Egypt and China, such commonly used patterns gradually developed into writing. The cuneiform writing of the Sumerians was impressed into clay tablets with wedge-shaped tools. (fig 59) The Egyptians' written signs, called *hieroglyphs* (fig. 60), were carved into stone or written on papyrus rolls. To the ancients, writing was a sacred art guarded by priests and scribes.

Fig. 59. Sumerian cuneiform medical text.

Fig. 60. Egyptian hieroglyphs from 12th dynasty (1991-1786 B.C.).

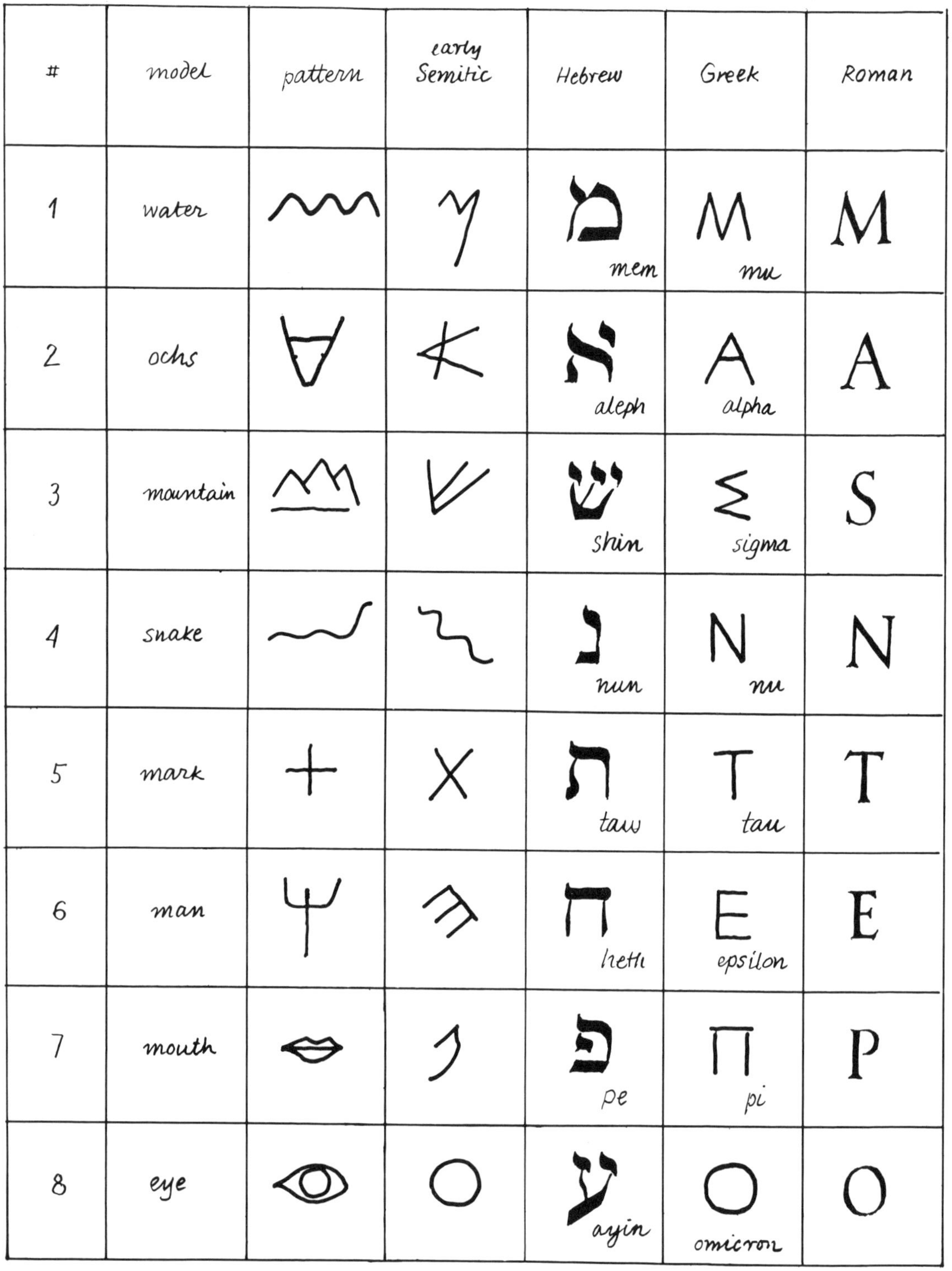

#	model	pattern	early Semitic	Hebrew	Greek	Roman
1	water	∿	ל	מ mem	Μ mu	M
2	ochs	∀	ⴲ	א aleph	Λ alpha	A
3	mountain	⋀⋀	∨	ש shin	Σ sigma	S
4	snake	∼	∼	נ nun	Ν nu	N
5	mark	+	✕	ת tau	Τ tau	T
6	man	Ψ	∍	ח heth	Ε epsilon	E
7	mouth	👄	ϟ	פ pe	Π pi	P
8	eye	👁	Ο	ע ayin	Ο omicron	O

For us, writing has become so commonplace that it has lost its sacred character, as with dancing. We find it hard to imagine what our ancestors might have felt when they first experienced the power of writing. The anthropologist Gordon Childe has said: "The immortalization of a word in writing must have seemed a supernatural process; it was surely magical that a man long vanished from the land of the living could still speak from a clay tablet or a papyrus roll. Words thus spoken must possess a kind of mana."[28]

Our own letters have been traced back to Roman and Greek prototypes, which in turn originated from still earlier proto-semitic images representing similar sounds, as figure 61 illustrates.

Fig. 61. Origin of some letter patterns.

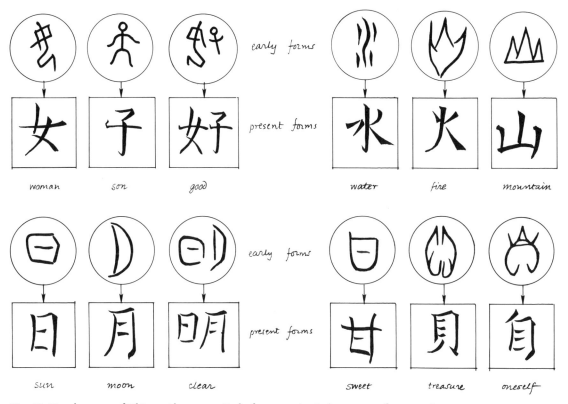

early forms

present forms

woman   son   good   water   fire   mountain

early forms

present forms

sun   moon   clear   sweet   treasure   oneself

Fig. 62. Development of Chinese ideograms. Early forms are in circles, current forms are in squares.

In China the patterns of writing never turned into signs of sounds as they did in the West. Instead, the signs of Chinese writing retain to this day the rudiments of the basic patterns from which they originated, with an idea attached to each, which is why they are called *ideograms*. In figure 62 a few Chinese ideograms are shown (in square frames), together with the basic image-patterns from which they originated (in round frames).

Even if one does not understand the meaning of a foreign written language, one can still share its rhythms, which reveal its unique character. The rhythm of Chinese writing—archaic as well as modern—moves with seemingly effortless grace (figs. 63, 64). The

Fig. 63. Chinese writing from bronze casting (11th century B.C.).

Fig. 64. Poem by Emperor Hui-tsung.

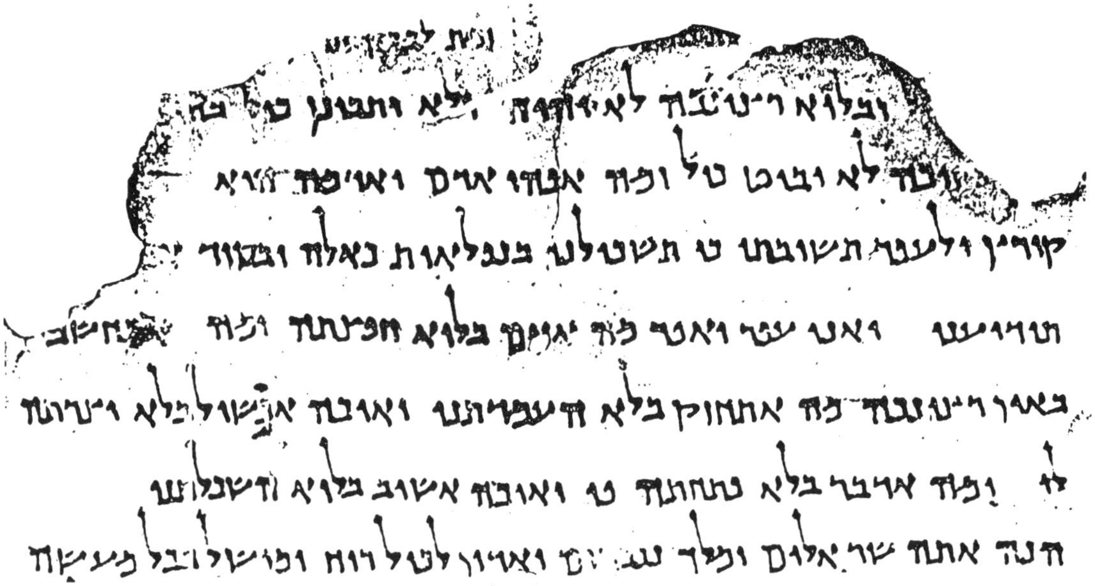

Fig. 65. Hebrew writing from Dead Sea Scrolls.

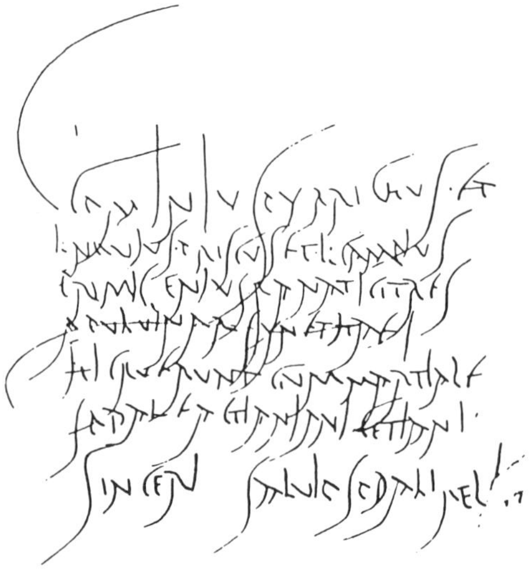

Fig. 66. Roman graffito from Pompeii.

Fig. 67. Arabic writing.

Hebrew writing of the Dead Sea scrolls (fig. 65) proceeds in a steadfast walk, as if it confirmed the prophet's words, "What doth the Lord require of thee but to do justly and . . . to walk humbly with thy God?"[29] In the Roman graffito scratched into a Pompeian wall (fig. 66) we recognize a familiar nervousness that reminds us of our own hurried longhand. The rhythm of Arabic dances with dignity (fig. 67), while the rhythm of one of the first Gutenberg bibles storms heavenward like the steeples of the Gothic town where it was printed (fig. 68).

The rhythms of writing are created by the same pattern-forming process of sharing that creates rhythms of dance, music and speech. Movements shared make dance, patterns shared make writing, and sounds shared make music and speech. It is also through a sharing process that we comprehend numbers. The fact that we all have ten fingers allows us to count the first ten numbers on our two hands. From ten, numbering becomes a rhythmically recurrent process: the tens make hundreds, the hundreds make thousands, and so on.

The ancients used the proportions of the human body for measuring short distances. For instance, the length of a man's forearm with the hand outstretched was called a *cubit,* which had

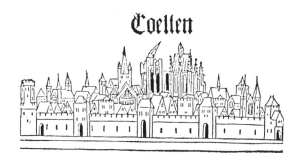

Coellen

obtulit holocaufta fup altare. Odu
ratufq; eft dns odore fuauitarif': z ait
ad eu. Mequaqz ultra maledicam ire
propter hoies. Senfus eui et cogitacō
humani cordis in malū prona funt
ab adolefcecia fua. Mon igitur ultra
percucia omnem animam uiuentem
ficut feci cundtis diebz· terre fauentis· z
meffis· fcigus z eftus· eftas z hiemps·
nox et dies non requi efcent. VI I I
Benedixitq; deus noe z filijs eius·
z dixit ad eos. Crefcite z multipli
camini z replete terrā· z terror uefter ac
trauor fit fup cūcta aīalia terre· z fupi
omnes uolucres celi cū uniuerfis que
mouent fup terrā. Dnanes pifces ma
ris: manui ueftre trad iti funt· et omne

Fig. 68. Lines from Gutenberg Bible (1452-1456) and
view of Cologne in 1400.

Fig. 70. Greek metrological relief. Distance from
fingertip to fingertip is a *fathom*; footprint above
figure indicates basic measure.

several variants. The Egyptians had a smaller cubit, which consisted of six "handbreadths" and a larger one, the Royal Cubit, which was seven handbreadths. (fig. 69) The Egyptian "hand" was made up of the dimensions of four "fingers" or "digits." A further measure, the "fist," equaling one and a third handbreadths, was used by the Egyptian master-builders and craftsmen to establish the square grids used for the proportioning of their royal statuary. Many unfinished Egyptian sculptures have been found which have such grids, used by the artists and workmen. The rhythmic wave diagram in figure 69 reveals how the main proportions of this figure share the harmonious proportions approximating the root harmonies of music.

The Roman cubit consisted of two feet, each foot measuring twelve *unciae;* this is the word from which our word *inch* comes. Six feet made up the *fathom,* which corresponded to a man's outstretched arms, from fingertip to fingertip. (fig. 70)

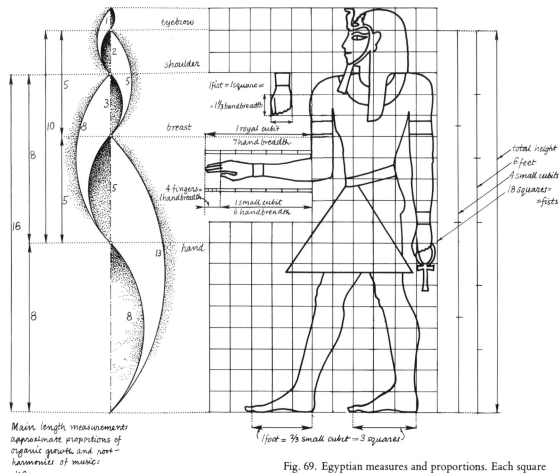

Main length measurements
approximate proportions of
organic growth and root-
harmonies of music:

1:2
5:10 = 0.5 = diapason-octave
8:16

2:3 = 0.666
3:5 = 0.6
5:8 = 0.625 $\Big\}$ = 0.618 = diapente = fifth.
8:13 = 0.615

Fig. 69. Egyptian measures and proportions. Each square of grid is a *fist,* corresponding to one-third of a foot.

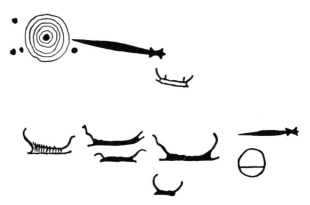

Fig. 71. Swedish rock engravings of ships and sunwheels (ca. 1300 B.C.).

Fig. 72. Swedish rock engravings of ships and swords (ca. 1300 B.C.).

As the measuring of length was linked to the proportions of the human body, so the measuring of time became based upon the durations observed in the rhymically occurring movements of the heavenly bodies. These observations led to the realization that there exists a heavenly or cosmic order, to which human life is related, and upon which it depends, this order manifesting itself in the movements of the stars and in the recurring patterns of the calendar.

Cosmic order and calendric structures

There are many indications that prehistoric people carefully observed the movements of heavenly bodies. In Scandinavian rock engravings from prehistoric times we see boats and sunwheels, with the spokes pointing in the directions of the compass. (fig. 71) There are also concentric circular sky patterns, one with a sword pointing towards it, apparently referring to a constellation which the early Vikings might have used as a navigational guide. (fig. 72) In another Swedish rock engraving, something like a giant's or a god's footprint is seen over the sunwheel, and below it is a figure with crescent-shaped headgear, perhaps a shaman, impersonating the "man in the moon," dancing in the flaming footsteps of the sun. (fig. 73)

Whatever the interpretation of such mythological patterns might be, they testify to early observations of the sky, which brought along better orientation in time and space.

Archeological and astronomical research has established that the large stone monuments which were built across Northern Europe about 3500 years ago served as giant compasses, calendars, and computers of seasonal patterns, as well as sacred precincts for religious rituals.[30] The most famous of these megaliths is Stonehenge, on the Salisbury Plain in England, built in stages from the twentieth to the sixteenth centuries B.C. (fig. 74)

A plan of Stonehenge III indicates how the exact time of midsummer sunrise was established (fig. 75), by sighting the disk of the rising sun between adjacent tall stone markers, called *Sarsen Stones,* directly on top of what is referred to as the *Heelstone.* (fig. 76) The horizontal projections (azimuthal directions) of the midwinter and midsummer risings and settings of the sun and the moon established with modern scientific instruments corroborate the astronomical accuracy of Stonehenge. (fig. 77)

Fig. 73. Swedish rock engravings of a deity or shaman, with footprint and sunwheel (ca. 1300 B.C.).

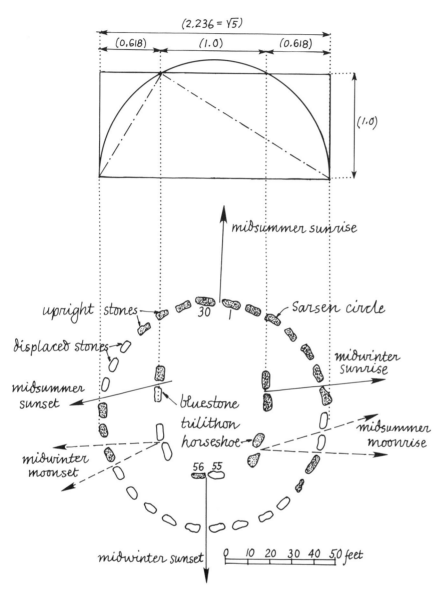

(2.236 = √5)

(0,618) (1.0) (0,618)

(1.0)

midsummer sunrise

upright stones

30

Sarsen circle

displaced stones

midwinter sunrise

midsummer sunset

bluestone trilithon horseshoe

midwinter moonset

midsummer moonrise

56 55

0 10 20 30 40 50 feet

midwinter sunset

Fig. 75. Alignments for Stonehenge III. Sarsen Circle's diameter and width of Bluestone Trilithon Horseshoe are in golden section relationship.

Fig. 74. Stonehenge.

Fig. 76. Heelstone framed in archway at Stonehenge.

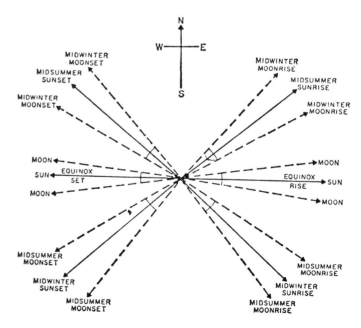

Fig. 77. Horizontal projections of sightlines (azimuthal directions) of rising and setting of sun and moon at solstice and equinox for the latitude of Stonehenge.

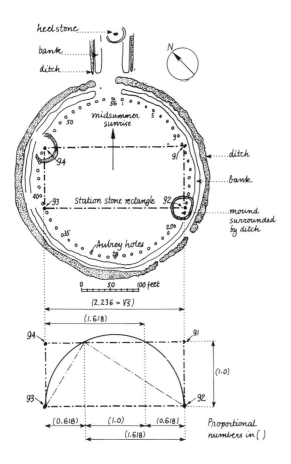

Fig. 78. Plan of Stonehenge I, *above,* with proportions of Station Stone Rectangle projected, *below,* showing its approximation of the √ 5 rectangle.

It seems to have escaped notice so far that the architecture of Stonehenge shares the proportions of the golden section and of the Pythagorean triangle. The classical construction of the golden section applied to plan III of Stonehenge (fig. 75) reveals that a golden relationship exists between the width of the Bluestone Trilithon Horseshoe and the diameter of the Sarsen Circle, (1:0.618 = 1.618). The same proportional construction applied to the plan of stage I shows that the rectangle formed by the Station Stones approximates the $\sqrt{5}$ rectangle, made up of two reciprocal golden rectangles. (fig. 78) The astronomer Gerald S. Hawkins, upon whose investigations the present analysis is based, believes that the Station Stone Rectangle was "immensely significant . . . historically, geometrically, ritualistically and astronomically."[31]

Geometric analysis with the help of center lines of piers and diagonals also shows that the proportions of the Sarsen archways come as close to the relationships of the 3-4-5 triangle as figure 79 indicates. Recalling that the 3:4 proportion corresponds to the musical harmony of diatessaron, the golden section corresponds to the diapente, and the $\sqrt{5}$ corresponds to the diapason, the phrase "music of the spheres" takes on new meaning.

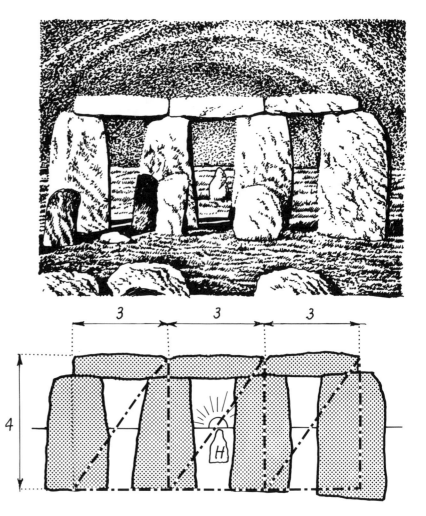

Fig. 79. Sarsen archways framing the Heelstone (H) at the tip of which the disc of the sun appears on the first day of summer. Proportions of archways are 3-4-5 triangles.

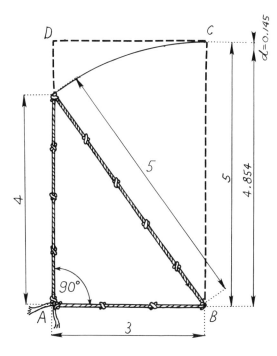

Fig. 81. Relationship between 3-4-5 triangle (ropestretcher's triangle) and golden rectangle.

Some of the same harmonious proportions can be found in a structure built about 1000 years before Stonehenge—the Great Pyramid of Egypt. Figure 80 shows that the central height (apothem) of any side triangle relates to half the base in the proportional ratio of the golden section. In cubits: ½ base = 220; apothem = 356; 356:220 = 1.618.[32]

As may be recalled, the 3-4-5-triangle also approximates the golden ratio between its 5-unit side and the 3-unit base. (5:3 = 1.666...). For the Egyptians, the 3-4-5 triangle was of vital importance. Their fields had to be surveyed yearly, because of inundations by the Nile, and this triangle served as their surveying tool. A rope with knots at twelve equal distances, held at the third and eighth knots with the ends brought together, produced the angle needed for surveying. This is why the 3-4-5 triangle is also referred to as the *ropestretcher's triangle* or the *Egyptian triangle*. The close relationship between the ropestretcher's triangle and the golden rectangle is illustrated in figure 81.

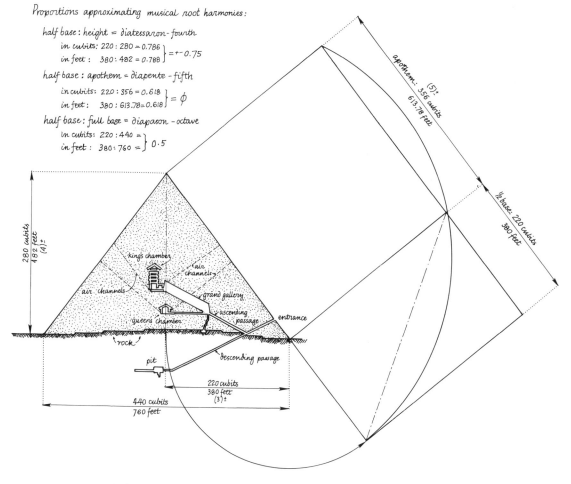

Fig. 80. Great Pyramid of Cheops at Gizeh. Cross section shows that apothem and half the base are in golden section relationship.

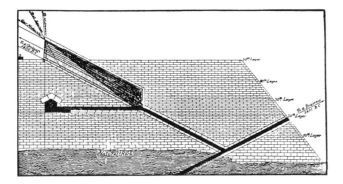

Fig. 82. Passages in the Great Pyramid, perhaps used as observation slots to measure the sun's height on the shortest and longest days of the year.

Fig. 83. Pyramid of the Moon, *above*, and the Pyramid of the Sun, *below*, Teotihuacan, Mexico.

Explorations have demonstrated that the Great Pyramid was not only a royal tomb, but was also a giant almanac "by means of which the length of the year, including its awkward 0.2422 fraction of a day could be measured as accurately as with a modern telescope . . . It also has been shown to be a theodolite, an instrument for surveying, that is precise, simple, and virtually indestructible."[33] As a compass, it is so finely oriented that modern compasses are adjusted to it, not vice versa. The astronomer Richard A. Proctor estimated, at the turn of the century, that the sun's height at its equinoxes as well as certain key stars and constellations could be sighted and aligned very accurately through the huge, graduated slots which were built into the body of the pyramid. Proctor speculates that the truncated top surface of the pyramid before its completion could have been used as an astronomical and astrological observation platform high over the Gizeh plateau, with a view commanding the entire compass.[34] (fig. 82)

In Mexico around 300 A.D., similar huge pyramidal structures were built to serve both religious and astronomical purposes, and they display basic proportional patterns similar to the Great Pyramid of Egypt and Stonehenge.

The Pyramids of the Sun and the Moon in Teotihuacan near Mexico City once formed the heart of a splendid metropolitan civilization. (fig. 83) Figure 84 shows how the outlines of the Pyramid of the Sun are contained within two sets of 5:8 and 3:4 triangles. Other

Fig. 84. Elevation and proportional analysis of the Pyramid of the Sun.

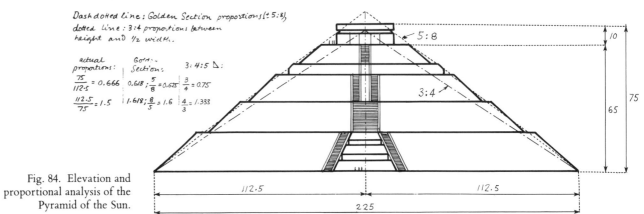

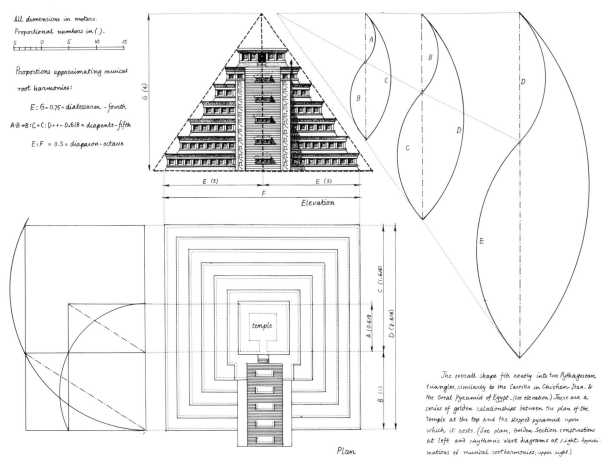

All dimensions in meters.
Proportional numbers in ( ).

Proportions approximating musical root harmonies:

E : G = 0.75 = diatessaron - fourth

A:B = B:C = C:D = + - 0.618 = diapente - fifth

E : F = 0.5 = diapason - octave

Elevation

Plan

The overall shape fits neatly into two Pythagorean triangles, similarly to the Castillo in Chichen Itza. & the Great Pyramid of Egypt. (see elevation.) There are a series of golden relationships between the plan of the temple at the top and the stepped pyramid upon which it rests. (See plan, Golden Section constructions at left and rhythmic wave diagrams at right. Approximations of musical root harmonies; upper right.)

Fig. 85. El Tajin: Pyramid of the Niches. Elevation and plan.

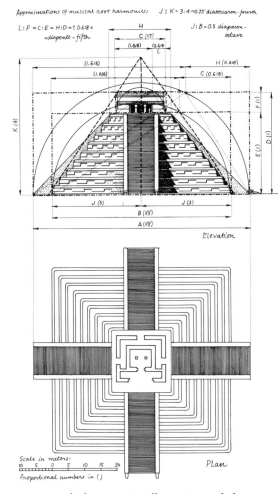

Approximations of musical root harmonies: J : K = 3 : 4 = 0.75 diatessaron - fourth

L : F = C : E = H : D = ± 0.618 =
= diapente - fifth

J : B = 0.5 diapason - octave

Elevation

Plan

Scale in meters:
Proportional numbers in ( )

Fig. 86. Chichen Itza: Castillo. Section and plan.

Mexican pyramids also fit into the 3:4 triangle's outlines, as can be seen from drawings representing elevations of two of them: the Pyramid of Niches in El Tajin, and Castillo in Chichen Itza. (figs. 85, 86) In addition, in El Tajin the golden section's proportions are approximated by a progressive series of relationships between cardinal dimensions, as indicated by the golden section constructions and wave diagram. In the Castillo, the main elements of the structure fit into a series of $\sqrt{5}$ rectangles (composed of two reciprocal golden rectangles) as the constructions superimposed over the elevation demonstrate. All these proportional relationships echo the root harmonies of music. The architect Manuel Amabilis Dominguez in 1930 discovered a prevalence of pentagonal proportions, and thus of the golden section, in many Mexican monuments.[35] Volumes have been written about the calendric significance of these pyramids. Apparently, in certain cases, even the number of structural or decorative elements correspond to the number of days in certain important cycles of time, and each stone had calendrical importance: in the Castillo the number of steps, and in the Tajin the number of niches, is 365, the number of days in the solar year.

Fig. 87. Ur Ziggurat ruins viewed from northeast.

Fig. 88. Drawing of Ur Nammu's
ziggurat after Sir Leonard
Woolley's theory.

...„and behold a ladder set upon the earth..."

Fig. 89. *The Tower of Babel* by Pieter Brueghel the Elder (1565).

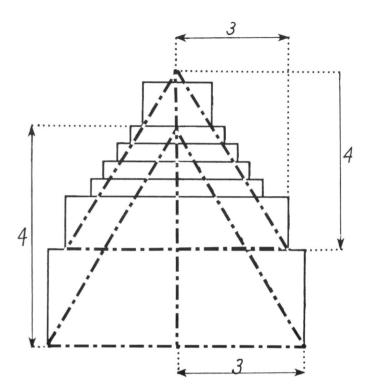

Fig. 90. Ziggurat of Babylon. Reconstruction by L.C. Stecchini.

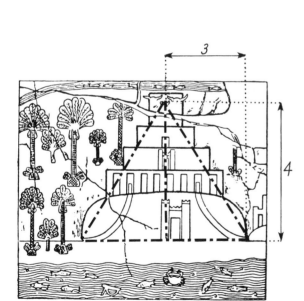

Fig. 91. Assyrian bas relief of ziggurat (7th century B.C.).

Observation of the heavens for clues to seasonal changes was also a major function of some of the world's oldest calendrical structures, the Ziggurats of Mesopotamia. According to Sir Leonard Woolley, who excavated the large Ziggurat of Ur, built by the Sumerians near today's Baghdad, the word *ziggurat* meant "hill of heaven" or "mountain of god." (fig. 87) Sir Leonard, whose reconstruction of the Ur Ziggurat is shown in figure 88, suggests that these huge stairways with processionals ascending and descending them might have had to do with Jacob's famous dream in the Bible: "and behold a ladder set upon the earth, and the top of it reached to heaven, and behold the angels of God ascending and descending on it."[36]

The priests of the ziggurats were also astrologers and astronomers who calculated the movements of planets, the sun and the moon, and established a lunar calendar for predicting seasonal changes, floods, sowing and harvesting times, and so on.

The most famous ziggurat was the Tower of Babel in Babylon, with its legendary hanging gardens of Semiramis. (fig. 89) Livio Catullo Stecchini reconstructed the basic outlines of the Ziggurat of Babylon on the basis of a cuneiform text, the so-called Smith Tablet.[37] The outlines of this reconstruction are shown in figure 90. Constructions of the golden section and Pythagorean triangle reveal correspondences to the fourth-diatessaron and fifth-diapente root harmonies of music. An Assyrian bas relief from the seventh century B.C. shows similar proportions. (fig. 91)

The plan, which is asymmetrical, shows tendencies to approximate the proportions of the Golden Section and of the Pythagorean triangle, as the 5:8 and 3:4 diagonals indicate. Detailed proportional relationships between neighboring parts and their multiples are tabulated numerically and graphically, showing that all proportional ratios tend to share the narrow band between the Golden Section's 0.62 ratio and the 0.75 ratio of the Pythagorean triangle.

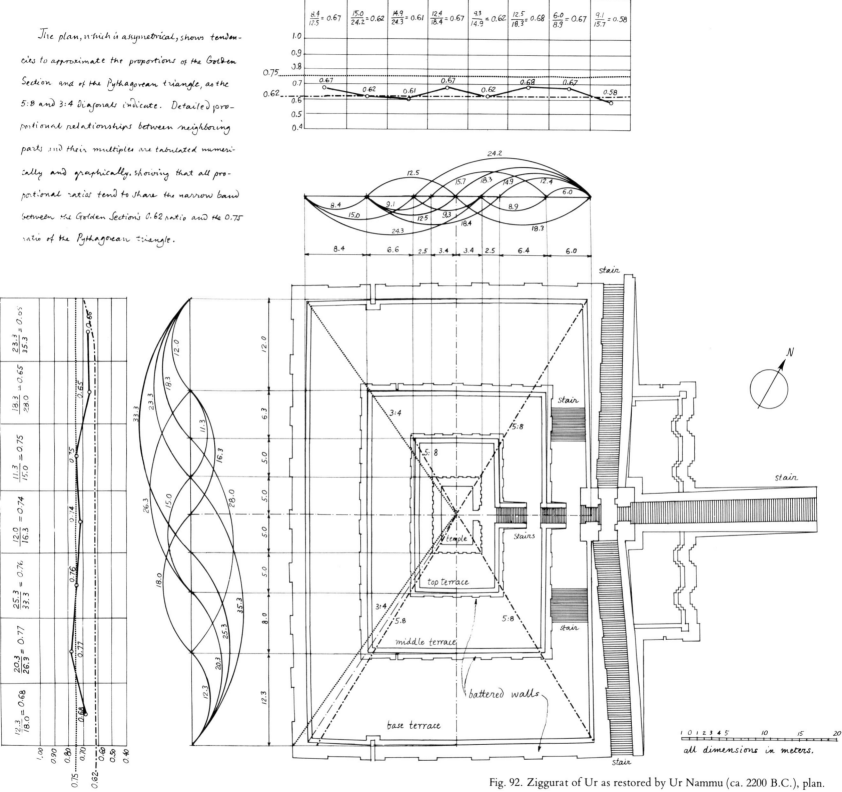

Fig. 92. Ziggurat of Ur as restored by Ur Nammu (ca. 2200 B.C.), plan.

Figures 92 and 93 are the author's proportional studies of the Ur Ziggurat, based upon Sir Leonard Woolley's excavations and of King Ur Nammu's reconstruction from around the twenty-second century B.C. (The age of the earlier ziggurats buried below the present ruins can only be guessed at; they could have been built as much as 1000 years earlier.) These proportional analyses indicate that the terraces and temple which make up this great structure share the proportions of the golden section and the Pythagorean triangle.

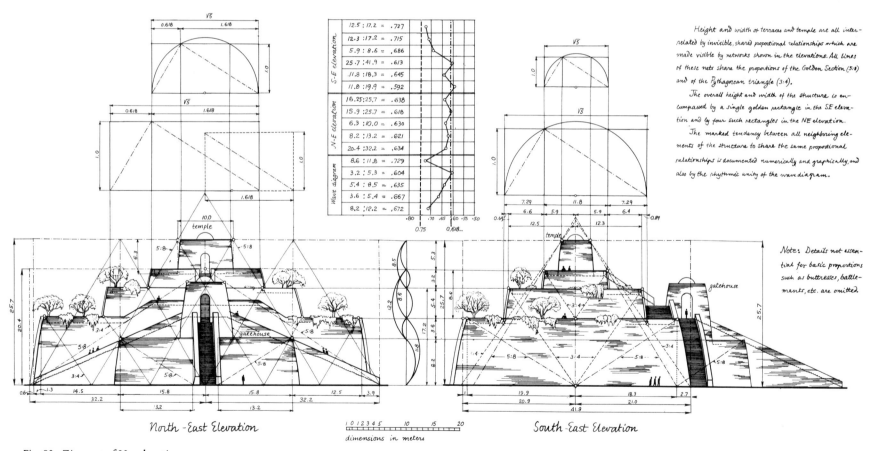

Fig. 93. Ziggurat of Ur, elevations.

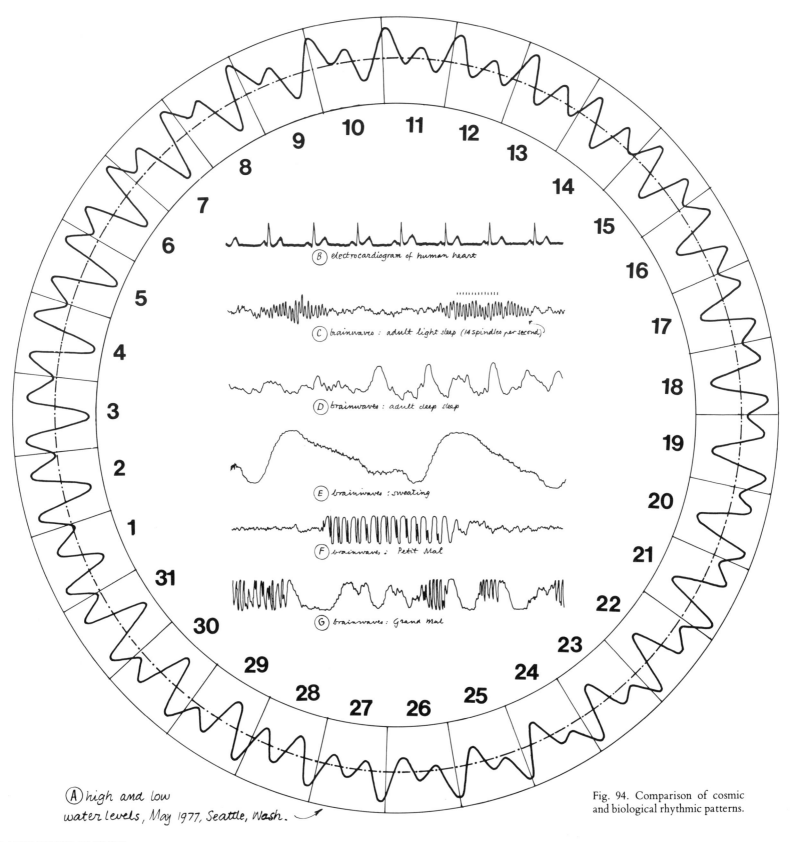

(B) electrocardiogram of human heart

(C) brainwaves: adult light sleep (14 spindles per second)

(D) brainwaves: adult deep sleep

(E) brainwaves: sweating

(F) brainwaves: Petit Mal

(G) brainwaves: Grand Mal

(A) high and low water levels, May 1977, Seattle, Wash.

Fig. 94. Comparison of cosmic and biological rhythmic patterns.

Rhythm and harmonious sharing

Calendric architecture revealed rhythmic patterns characteristic of the calendric changes themselves—the wanings and waxings of the moon, the rhythms of ebbs and tides, or menstrual cycles of the female body.

The circular pattern in figure 94 represents a page of the calendar for May 1976. Every square represents a day's record of the ebbs and tides of the Pacific Ocean in Seattle, Washington. Fluctuations within the major rhythm are caused by the joint gravitational pull of the moon and the sun, the former being the stronger one because of its relative closeness to the earth.

On a miniature scale, the wave patterns of this cosmic rhythm are shared by our heartbeat, as electrocardiogram **B** in figure 94 indicates. Our brain waves are further variations of these rhythms, depending upon our condition—light sleep, deep sleep, sweating, or the mental disturbances called *petit mal* and *grand mal,* for instance.

Our breathing has a similarly dinergic wave pattern: it is an ongoing rhythm of inhaling and exhaling, and similar fluctuations characterize the physical and mental cycles called *biorhythms.* Our "inner clocks" allow us to register our own rhythmic patterns of time, called *circadian* rhythms. Air travelers become aware of their circadian rhythms when crossing time zones: their own inner time and outer, calendric time temporarily get out of step.

By attaching pens to a pendulum and to a tuning fork, and by registering the pens' traces on rolling paper, the mathematical physicist Sir James Jeans proved that weight and sound share the same rhythmical, harmonious wave patterns, called *sine curves,* or *simple harmonical curves.*[38] (fig. 95)

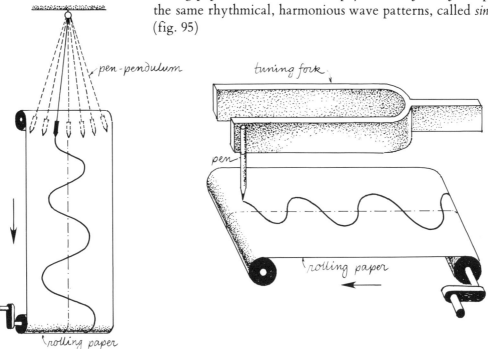

Fig. 95. Simple harmonic patterns created by pendulum, *left,* and tuning fork, *right.*

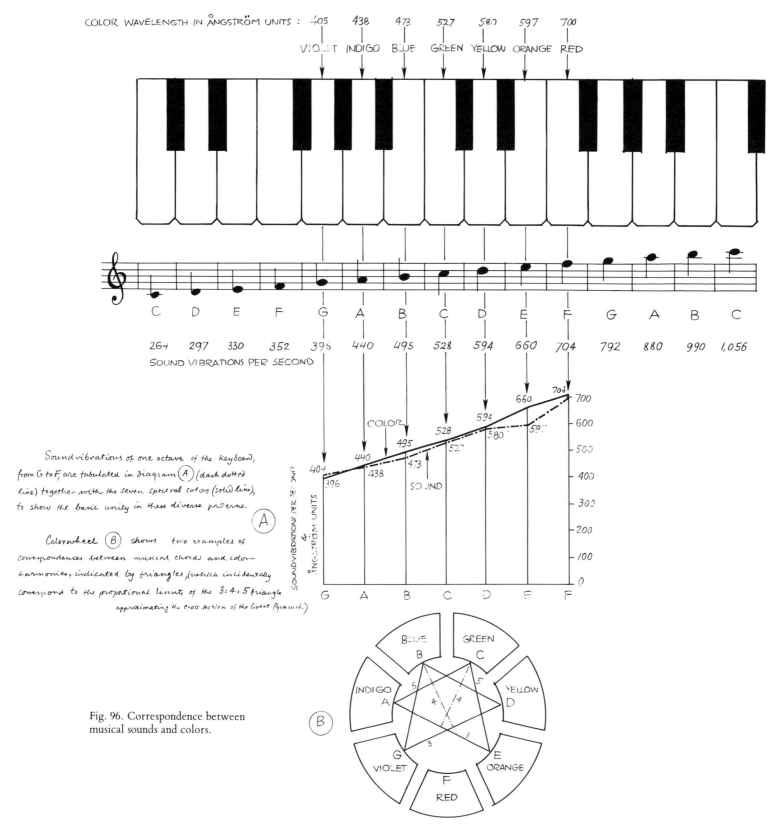

COLOR WAVELENGTH IN ÅNGSTRÖM UNITS : 405   438   473   527   580   597   700

VIOLET INDIGO BLUE GREEN YELLOW ORANGE RED

C   D   E   F   G   A   B   C   D   E   F   G   A   B   C

264   297   330   352   396   440   495   528   594   660   704   792   880   990   1,056

SOUND VIBRATIONS PER SECOND

Sound vibrations of one octave of the keyboard, from G to F, are tabulated in diagram Ⓐ (dash dotted line) together with the seven spectral colors (solid line), to show the basic unity in these diverse patterns.

Colorwheel Ⓑ shows two examples of correspondences between musical chords and color-harmonies, indicated by triangles, (which incidentally correspond to the proportional limits of the 3:4:5 triangle approximating the cross section of the Great Pyramid.)

Fig. 96. Correspondence between musical sounds and colors.

Light, color and sound also share the same wave patterns. What is more, they share the same vibration rates, as J. Dauven proved in 1970 (see figure 96.)[39] Diagram **A** is a composite of both vibrations; sound frequencies (the number of vibrations per second) are shown with a dash-dotted line, and color frequencies are shown with a solid line. The closeness of the two lines indicates that the experience of harmonious rhythms is shared by the eye and the ear, even though one registers it as color, the other as sound. The shared harmonies of colors and musical chords are shown in diagram **B.** Here the double 3-4-5 triangles indicate the tonic chord A minor (**A-C-E**), corresponding to indigo-green-orange on the color wheel, which might be seen when an orange tree is in fruit and the leaves cast dark blue shadows. The tonic chord G major (**G-B-D**) corresponds to violet-blue-yellow, the colors of an iris or violet under a sunny blue sky.

The essence of all vibration and rhythm is a sharing of diversities—weak and strong, in and out, up and down, back and forth—at recurrent time intervals. This holds as true for the tides of the ocean as for our heartbeat; for light, weight and sound as for patterns of plant growth.

If we compare our reconstructions of leaf outlines with so-called *isocandle* light distribution patterns of lighting fixtures, as in figure 97, we see definite correspondences. The intensities of light are measured and charted in these diagrams along radiating rays emanating in straight lines from the center of the fixtures, spreading in ever wider orbits, represented by concentric circles. This pattern is thus as dinergic in origin as the growth pattern of leaves or flowers, created by radiational and rotational growth components.

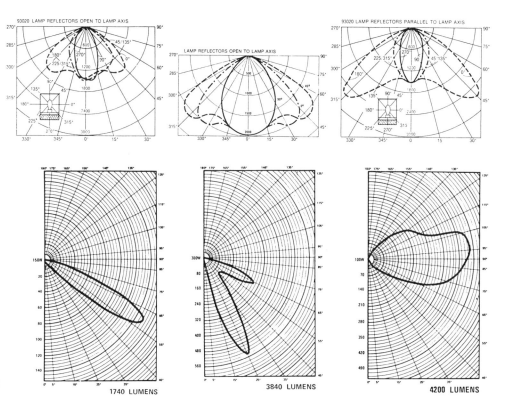

Fig. 97. Isocandle diagrams showing light distribution patterns of various light fixtures.

A row of sunflower seeds along spiral **D** in figure 7 shows that the diagonal length of the seeds group themselves in harmonical series represented by vertical bars in diagram **2** of figure 98. This pattern is similar to that of the lilac leaf (fig. 17) and the cumulative length of neighboring groups of seeds—represented by the upper and the lower bars. All share the same golden 5:8 relationship, approximating the musical root harmony of fifth-diapente. (fig. 98, diagram 3)

Counting the number of seeds in neighboring groups, as shown by the bars, we get:

Y-E	Z-F	H-G	K-J	M-L	O-N	
$\dfrac{5}{5}$	$\dfrac{5}{6}$	$\dfrac{5}{7}$	$\dfrac{5}{8}$	$\dfrac{5}{9}$	$\dfrac{5}{10}$	etc. Divided by 5 =
$\dfrac{1}{1}$	$\dfrac{1}{1.2}$	$\dfrac{1}{1.4}$	$\dfrac{1}{1.6}$	$\dfrac{1}{1.8}$	$\dfrac{1}{2}$	etc.

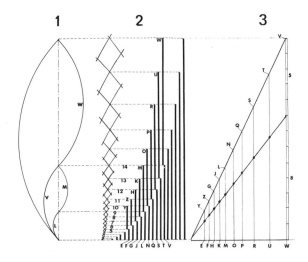

Fig. 98. Diagrams of a row of sunflower seeds, along spiral **D** in fig. 7. **1.** Rhythm of golden relations between neighboring groups of seeds; **2.** diagram of harmonic seed series; **3.** prevalence of golden section proportions.

Such a series is referred to in algebra as a *harmonic progression,* a concept that plays an important role, for instance, in the mathematical interpretation of sounds. A harmonic progression is defined as a series of fractions in which the nominator remains 1 while all neighboring denominators share the same difference, in this case 0.2. Put another way, a harmonic progression consists of the reciprocals of an arithmetical series. Recalling that in the Fibonacci series the difference between neighboring numbers also approximates the same ratio—ϕ moving one way and $1/\phi$ moving the other way—we see that the Fibonacci series is a truly harmonic progression.

These few random examples illustrate that all rhythmic vibration is essentially harmonious sharing. Since this sharing is universally present in musical sound, color, light and weight, patterns of plant growth, ebbs, tides and calendric rhythms, as well as in our own biorhythms, breathing, and heartbeat, we can speak of it as a basic pattern-forming process. In the next chapter we will look into the role this pattern-forming process plays in shaping the anatomy of animals.

CHAPTER 5: The Anatomy of Sharing

Shells, clams, crabs and fishes

The shapes of shells have been the subject of many studies which show that their harmonious shapes unfold in logarithmic spirals characterized by the golden section's proportions. A logarithmic spiral typical of shell growth, figure 99, shows that each consecutive stage of growth is encompassed by a golden rectangle which is by a square larger than the previous one, a pattern which Jay Hambidge called the pattern of "whirling squares."[40] One can trace this "whirl" by following the table below:

golden rectangles	+ squares	= golden rectangles:
0 1 2 3	+ 0 3 4 5	= 1 2 4 5
1 2 4 5	+ 1 5 6 7	= 2 4 6 7
2 4 6 7	+ 2 7 8 9	= 4 6 8 9
4 6 8 9	+ 4 9 10 11	= 6 8 10 11 etc.

The harmony of the Fibonacci series is present in this shell curvature, as the equidistant radii in figure 100 illustrate.

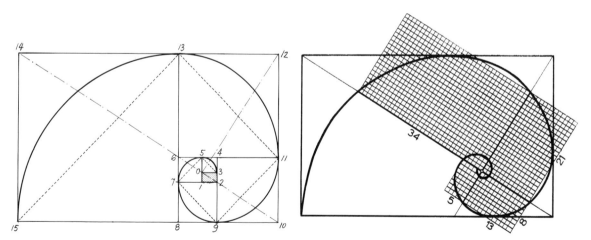

Fig. 99. Logarithmic spiral, typical of shell growth. Successive stages of growth are marked by "whirling squares" and golden rectangles growing in harmonic progression from center **O** outward.

Fig. 100. Fibonacci numbers in logarithmic spirals, typical for shells.

Fig. 101. Atlantic sundial
(*Architectonica nobilis*).

That even widely differing shell shapes share the proportions of golden dinergy can be seen by comparing, for instance, the Atlantic sundial (*Architectonica nobilis*) (fig. 101)— which is almost perfectly circular—with a delicate variety of abalone (*Haliotis asinina*) (fig. 102) which is elongated like a donkey's ear; (asininus means "donkey" in Latin). Successive stages of growth in the abalone—measured along neighboring, equidistant radii—are Fibonacci numbers. (fig. 103)

Fig. 102. Abalone (*Haliotis asinina*). The shape above the dotted line is reconstructed in Fig. 103.

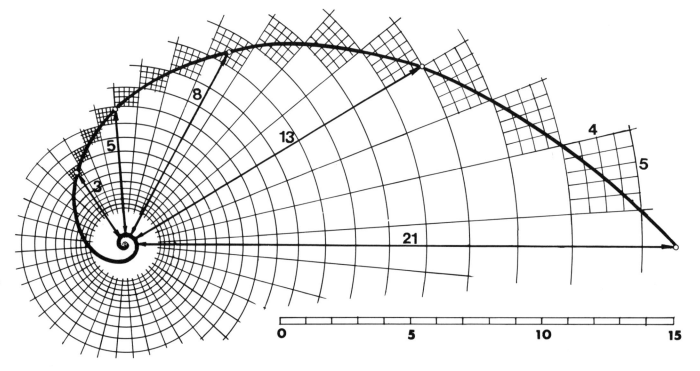

Fig. 103. Dinergic reconstruction of abalone shell's outline.

In the Atlantic sundial, because of its nearly circular shape, it is easy to measure successive increments of growth projected along one of the radii, marked **a** to **l** in figure 104. The reconstructions—by the same dinergic method of combined radial and rotational growth increments as used earlier for the reconstructions of daisy, sunflower and leaf outlines—reveal that the difference between the curvatures is only in the tempo, or rate of unfolding. The sundial's spiral moves with relative slowness between consecutive circles. It crawls through not less than 20 radii, or squares, before reaching from one circle to the next: a proportion of 20:1 between radial and rotational growth. The abalone takes only four to five radii or squares to move from one circle to the next: a proportion of 4 or 5:1. Constructions of the golden section at the right of figure 104 prove that the proportions of neighboring whorl-widths are the golden ones: all the shaded rectangles—width and length of which correspond to the width of neighboring whorls—are golden rectangles. The harmony created by this relationship is illustrated also by the organ-pipe-like bars, indicating the cumulative length of successively widening whorls. This sharing of the same proportional limitations throughout the entire growth process is also expressed by a series of equations all equaling ϕ.

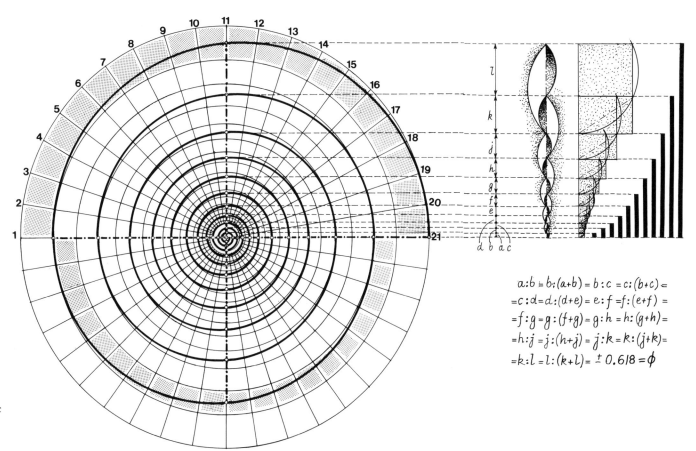

$$a:b = b:(a+b) = b:c = c:(b+c) =$$
$$= c:d = d:(d+e) = e:f = f:(e+f) =$$
$$= f:g = g:(f+g) = g:h = h:(g+h) =$$
$$= h:j = j:(h+j) = j:k = k:(j+k) =$$
$$= k:l = l:(k+l) = \pm 0.618 = \phi$$

Fig. 104. Dinergic reconstruction of Atlantic sundial.

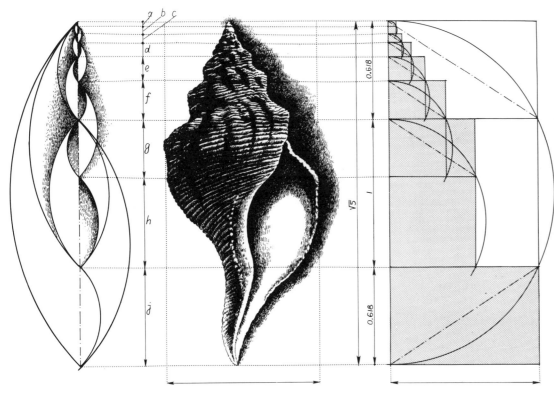

$a{:}b = b{:}(b{+}c) = b{:}c = c{:}(c{+}d) = d{:}e = e{:}(d{+}e) = e{:}f = f{:}(e{+}f) = f{:}g = g{:}(f{+}g) = g{:}h = h{:}(g{+}h) = j{:}(g{+}h) = (g{+}h){:}(g{+}h{+}j) = (a{+}b{+}c{+}d{+}e{+}f){:}(g{+}h) = 0.618... = \varphi$

Fig. 105. Dilated whelk (*penion dilatus*).

Of course, shells grow in all three dimensions of space simultaneously. In the abalone, however, and in the sundial, development in the third dimension of depth is relatively small. In shells that have greater depth, this development also shares the proportions of golden dinergy, as can be seen, for instance, in the dilated whelk of New Zealand (*Penion dilatatus*), figure 105. The wave diagram at the left indicates how all neighboring whorls share the same golden relationships. The long equation series at the bottom of the drawing expresses the same thing. The right-hand diagram furnishes constructions of the golden section as in the previous example.

A side view of the bear's paw clam (fig. 106) illustrates the marvelous precision of dinergic sharing between the two opposite valves interlocking and complementing each other. Reconstructions of the basic outlines reveal that the harmony of this lovely shape shares the same golden dinergies as so many earlier examples of organic growth.

Fig. 106. Bear's paw clam (*Hippopus hippopus*). Dinergic reconstruction of central spiral, *upper diagram*, and cross section, *lower diagram*.

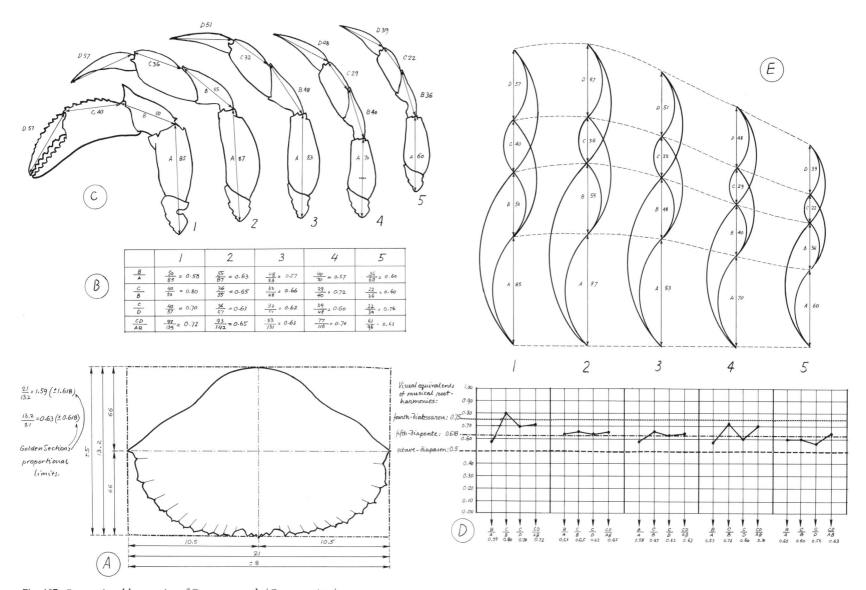

Fig. 107. Proportional harmonies of Dungenes crab (*Cancer magister*).

In crustaceans we again find golden dinergic relations shared by all neighboring parts of the anatomy. The carapace of a typical Dungeness crab, from the Northwest Pacific coast of the U.S., fits into a golden rectangle as drawing **A** of figure 107 shows. The length of neighboring parts of the pincers and legs (**C**) in relation to each other (table **B**) fluctuates between visual equivalents of the three musical root harmonies of 0.75, fourth-diatessaron; 0.618, fifth-diapente; and 0.5, octave-diapason (**D**). The wave diagrams of pincer and legs (**E**) show the harmonious rhythm created between all these differently sized members by their shared proportional relationships.

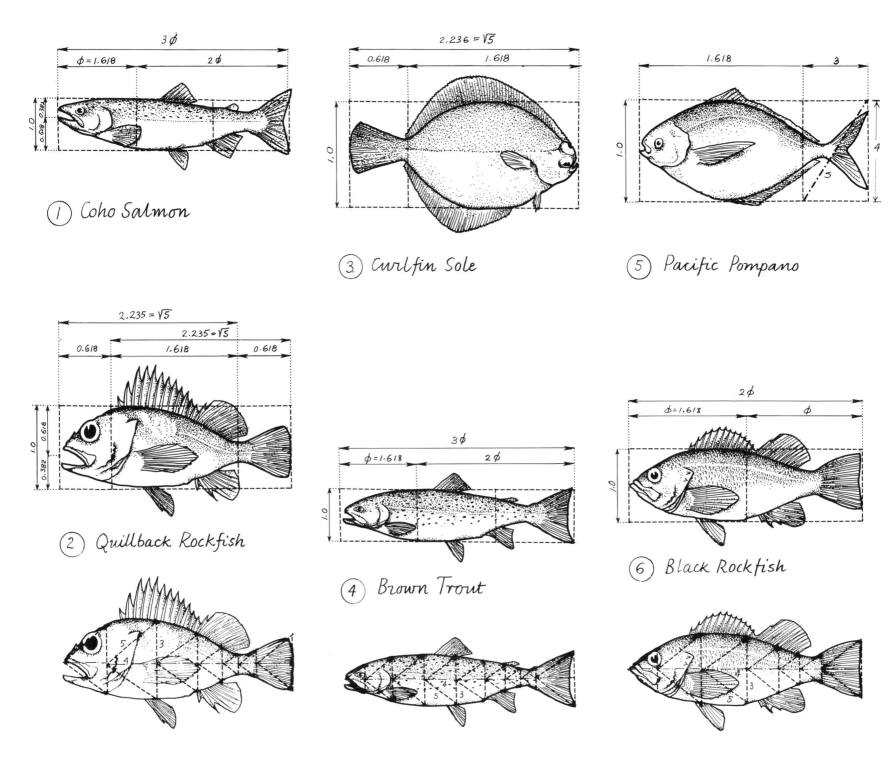

Fig. 108. Shared proportions of fish.

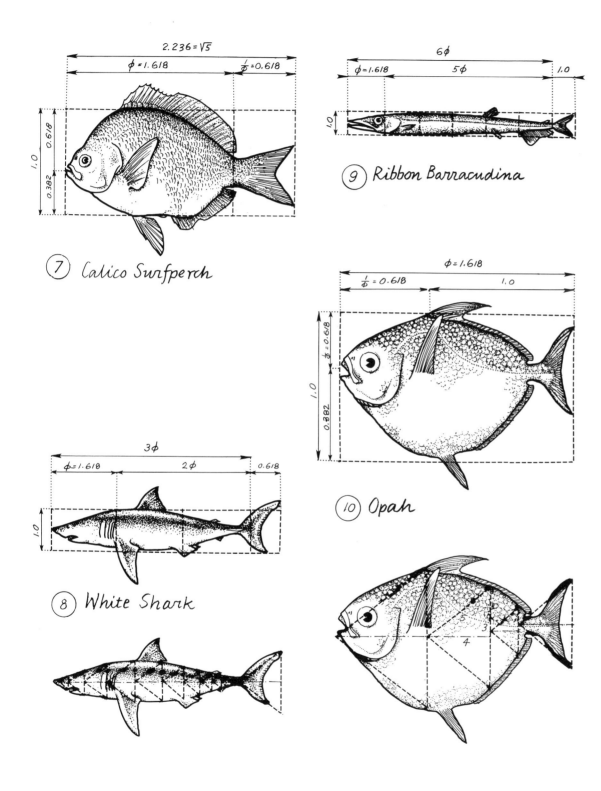

2.236 = √5

φ = 1.618

1/φ = 0.618

0.618

1.0

0.382

⑦ *Calico Surfperch*

6φ

φ = 1.618

5φ

1.0

1.0

⑨ *Ribbon Barracudina*

3φ

φ = 1.618

2φ

0.618

1.0

⑧ *White Shark*

φ = 1.618

1/φ = 0.618

1.0

1/φ = 0.618

1.0

0.382

⑩ *Opah*

3

4

A study of a variety of fish shapes reveals similar rhythmic harmonies resulting from similarly shared proportional limits. The proportional analysis of ten different, randomly chosen fish from the Pacific waters of Canada demonstrates that their basic outlines—and frequently also the detailed articulations of their bodies—share in a variety of ways both the proportions of the golden section and of the 3-4-5 triangle. The pictures of each of the ten fish in figure 108 illustrate how the outlines fit into golden rectangles, their multiples and reciprocals, at times combined with squares. In many instances, the mouth is at the golden section point of the body's height, as in the Coho salmon (**1**), quillback rockfish (**2**), calico surfperch (**7**), and opah (**10**). The lower row of fish shows how a series of double 3-4-5 triangles can be fitted into the outlines from mouth to tail.

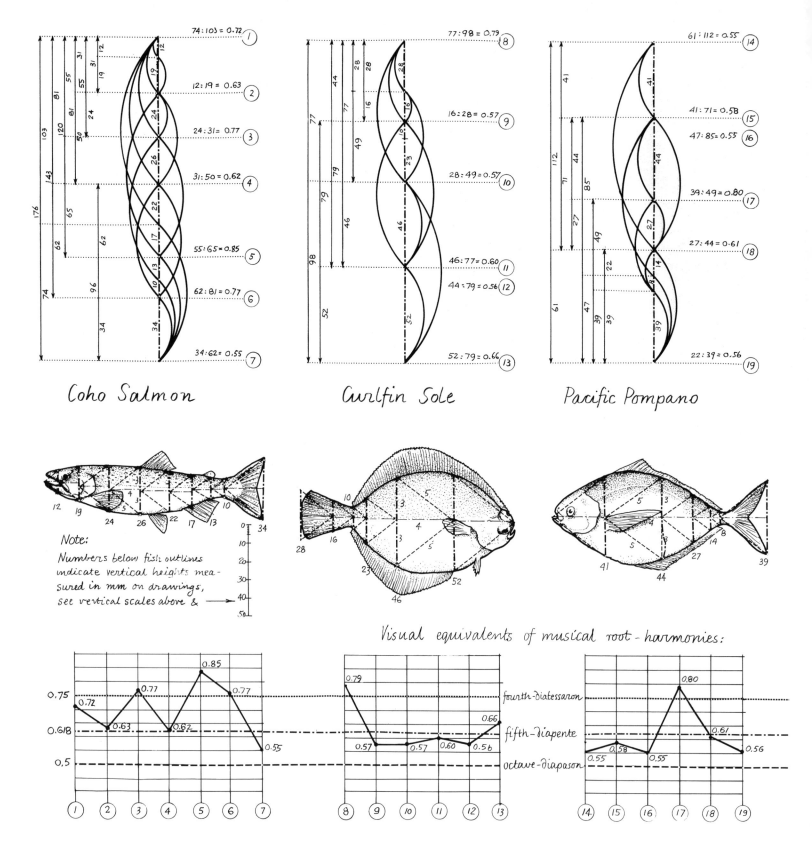

Coho Salmon

Curlfin Sole

Pacific Pompano

Note:

Numbers below fish outlines indicate vertical heights measured in mm on drawings, see vertical scales above &

Visual equivalents of musical root-harmonies:

fourth-diatessaron

fifth-diapente

octave-diapason

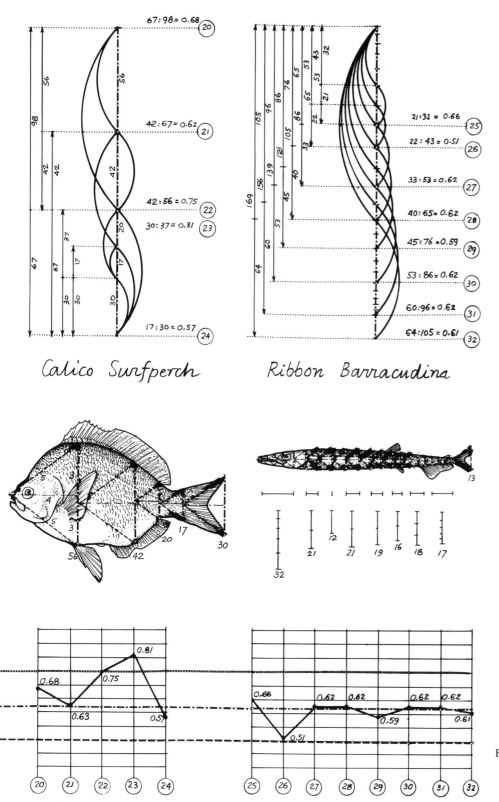

The combined three-unit length of such 3-4-5 triangulation diagrams, representing vertical body height, constitute true harmonic progressions, shown in figure 109 for the coho salmon, curlfin sole, Pacific pompano, calico surfperch, and ribbon barracudina. The harmonies of these rhythms likewise approximate the root harmonies of music, as wave diagrams and graphs show.

Calico Surfperch

Ribbon Barracudina

Fig. 109. Rhythmic harmonies of fish.

Even in the third dimension of thickness, fish tend to partake of the golden section's proportions, as three typical transverse sections in figure 110 illustrate.

The same proportional limits prevail in the shapes of skates and stingrays, as illustrated by figure 111. The main bodies of the deep-sea skate (**1**), black skate (**2**), longnose skate (**3**), and starry skate (**4**), are each encompassed by two golden rectangles, while the big skate's body (**5**) fits into a single golden rectangle. The greatest width of the bodies invariably tallies with the golden section point of the height.

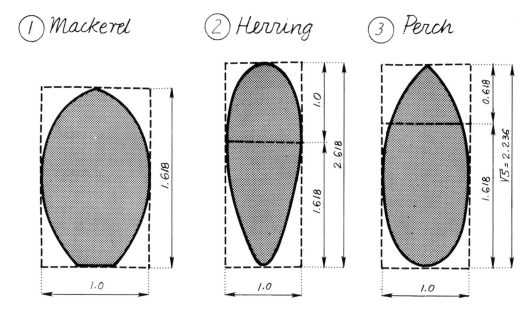

The mackerel's transverse section ① fits into a single golden rectangle. More compressed shapes like

the herring ② and the perch ③ share these proportions in the following ways: the herring adds its width to the

height of the golden rectangle which encloses the rest of its shape; the height of the perch is $\sqrt{5}$, which is the length of a golden rectangle and its reciprocal.

Fig. 110. Proportions of transverse sections of fish.

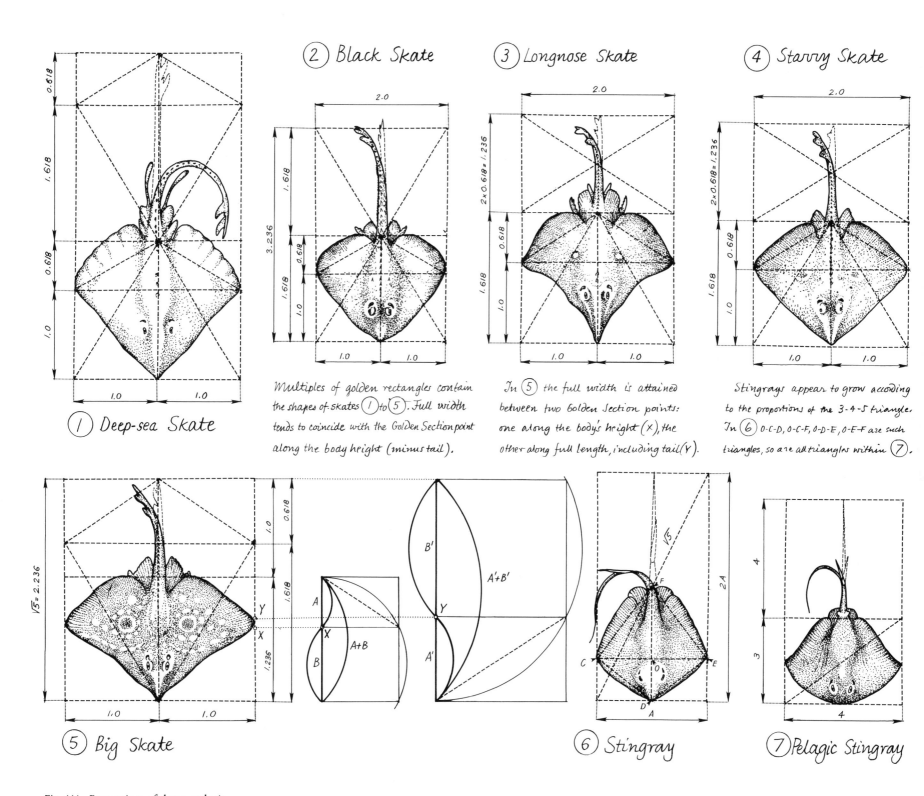

② Black Skate

③ Longnose Skate

④ Starry Skate

① Deep-sea Skate

Multiples of golden rectangles contain the shapes of skates ① to ⑤. Full width tends to coincide with the Golden Section point along the body height (minus tail).

In ⑤ the full width is attained between two Golden Section points: one along the body's height (X), the other along full length, including tail (Y).

Stingrays appear to grow according to the proportions of the 3-4-5 triangle. In ⑥ 0-C-D, 0-C-F, 0-D-E, 0-E-F are such triangles, so are all triangles within ⑦.

⑤ Big Skate

⑥ Stingray

⑦ Pelagic Stingray

Fig. 111. Proportions of skates and stingrays.

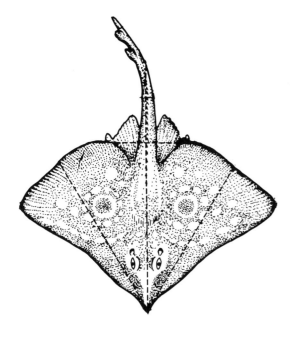

The length of the tail shares the same proportional limits in a variety of ways outlined in the drawings. In the deep-sea skate, the tail is as long as one of the golden rectangles encompassing the body plus a reciprocal totaling $\sqrt{5}$. In the black skate, the length of the tail and the height of the body are about equal. The tails of the longnose skate and the starry skate correspond to the shorter side of a golden rectangle, the longer side of which equals the width of the body. Two $\sqrt{5}$ rectangles encompass the entire shape of the big skate, from end of tail to tip of snout.

The two stingrays in figure 111 display a preference for the 3-4-5 triangle's proportional limits. The main body outlines of the stingray (**6**) are encompassed by two sets of reciprocal 3-4-5 triangles: **OCD, OCF, OED, OEF;** the full width corresponds to the joint side of these two sets of triangles and the length of the tail doubles the height of the body. The pelagic stingray's body (**7**) fits into a rectangle composed of two 3-4-5 triangles and its tail's length is the same as the body's width.

Let us now move up the evolutionary ladder from the denizens of the deep to the vertebrates of dry land, starting with a prehistoric giant reptile.

Dinosaur, frog and horse

The skeleton of the giant reptile allosaurus, a fierce carnivorous dinosaur that lived about 140 million years ago, was unearthed in Utah and recently was reassembled at the Thomas Burke Memorial Museum of the University of Washington, Seattle. Figure 112 shows the complete skeleton, which the author had the privilege to measure. As in earlier examples, a wave diagram illustrates the articulation of this huge body along the vertebral column into skull, neck, trunk, sacrum and tail corresponding to the musical root harmonies of fifth-diapente and fourth-diatessaron. Point **A,** where the sacrum and the first tail vertebra meet, is the golden section point along the full length of the skeleton, relating the tail to the rest of the body similarly as the trunk relates to skull plus neck, and neck relates to skull, the latter's length and height again sharing the same relationship. The tabulation shows how these relationships fluctuate around the 0.75 ratio of the Pythagorean triangle, and the golden section's 0.618 ratio. The length of sacrum and trunk are in a reciprocal golden relationship (within a fractional difference, marked **d**), as the construction of two reciprocal golden rectangles over dorsal and sacral vertebrae indicates.

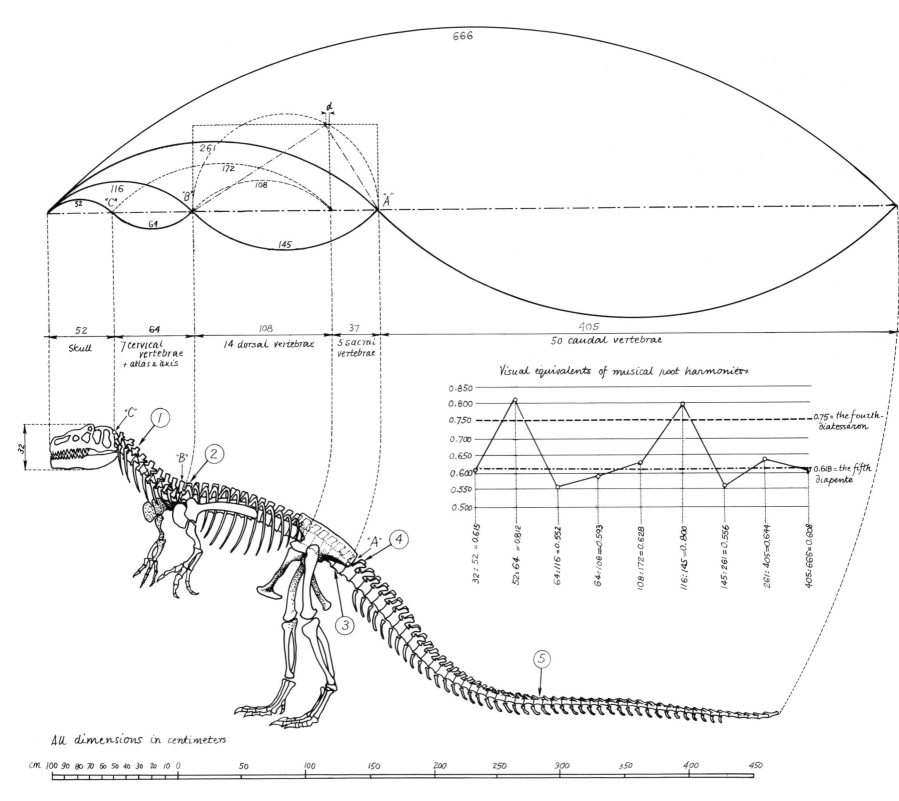

666

d

261

172

108

116

52 "C"

"B"

"A"

64

145

52
Skull

64
7 cervical
vertebrae
+ atlas & axis

108
14 dorsal vertebrae

37
5 sacral
vertebrae

405
50 caudal vertebrae

"C" ①

"B" ②

32

"A" ④

③

⑤

Visual equivalents of musical root harmonies

0.850
0.800
0.750
0.700
0.650
0.600
0.550
0.500

0.75 = the fourth-
diatessaron

0.618 = the fifth
diapente

32 : 52 = 0.615
52 : 64 = 0.812
64 : 116 = 0.552
64 : 108 = 0.593
108 : 172 = 0.628
116 : 145 = 0.800
145 : 261 = 0.556
261 : 405 = 0.644
405 : 666 = 0.608

All dimensions in centimeters

cm 100 90 80 70 60 50 40 30 20 10 0   50   100   150   200   250   300   350   400   450

Fig. 112. Dinergic reconstruction of allosaurus.

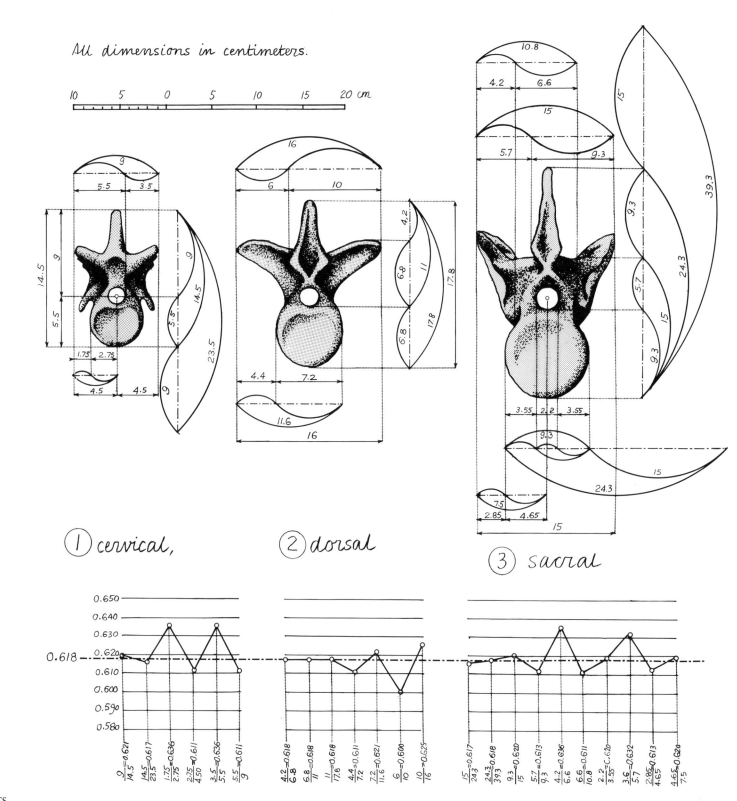

All dimensions in centimeters.

① cervical, ② dorsal ③ sacral

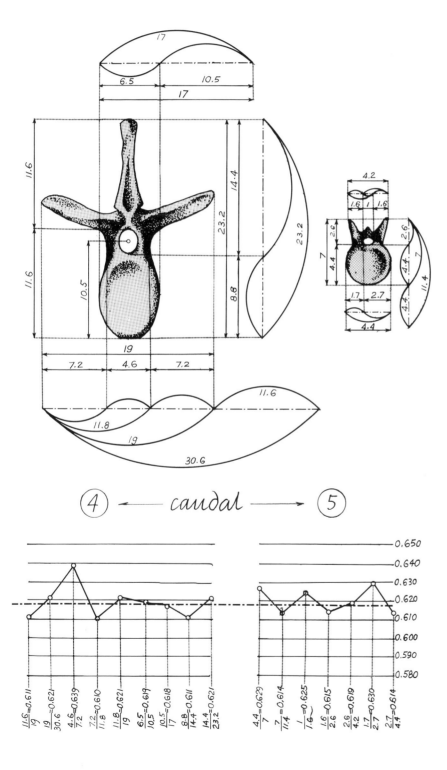

Every single vertebra shares the same proportional relationships that unite the whole body, and every one does it in a different way, appropriate to its station within the whole body and to its specific function, as shown by figure 113. Thus the sacral vertebrae (**3**) are the strongest and most massive since they act as fulcrum, balancing the enormous weights of trunk and head on one side and the huge tail on the other. The caudal vertebrae (**4** and **5**) are the most slender since they support only themselves. The laterally protruding process bones of the dorsal vertebrae (**2**) are much heavier than those of the tail, since they help to carry the weight of the trunk and the head. The cervical vertebrae (**1**) are the most complex ones: not only do they carry the weight of the huge head, but they also accommodate its turning in every direction.

The wave diagrams reveal that the rhythmic unity of these delicately sculptured shapes is based on a sharing of golden relations between many diverse neighboring parts. For instance, the round hole of the nerve channel is always very near the golden section point of the total height; the lateral protrusions of process bones and the vertical extension of the spine tend to share this relationship with the proportions of the bulky round bone at the base, called the *body*. The tabulation shows again that, while none of these relationships is mathematically perfect, the general tendency of approximations is unmistakable.

Fig. 113. Proportions of vertebrae of allosaurus.

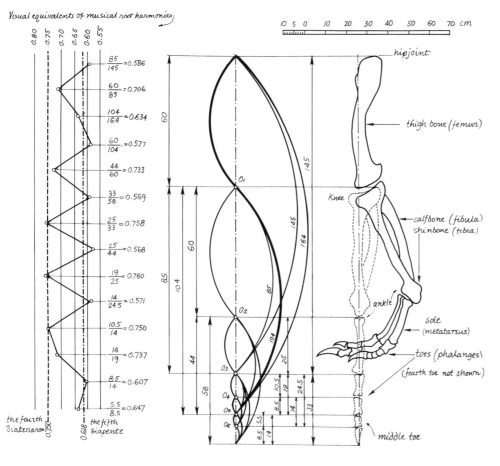

Details of hindleg and foot (fig. 114), as well as frontleg and foot (fig. 115), show the same kind of rhythmic sharing of the narrow range of proportional limitations between small and large bones as we observed within the vertebrae and the major articulations of the whole body.

All neighboring bones share proportional limitations corresponding to the musical root harmonies of the fourth and the fifth, as the graphs and wave diagrams indicate. Even the small coracoid bone at the shoulder shares golden proportions between its width and length. Furthermore, all joints—shoulder, elbow, wrist, hip, knee, ankle, as well as the minute joints of the phalanges in hands and feet—form harmonious series which also share these same proportional limitations.

Fig. 114. Proportions of hindleg and foot of allosaurus.

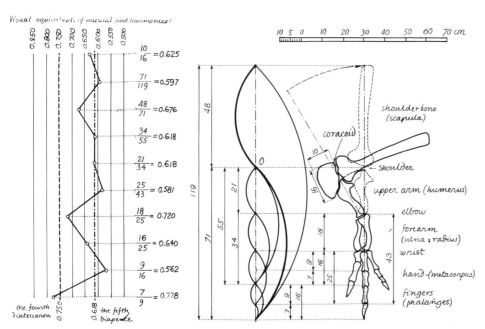

Fig. 115. Proportions of front leg and foot of allosaurus.

The three pairs of bones which constitute the pelvic girdle of the allosaurus (fig. 116) are of particular interest: each of these pairs is totally different from the others, each serving a different function, yet they too are shaped and related to each other by the same proportions that shape and unite the rest of the body. The illium fits into a single golden rectangle, the ischium into two golden rectangles, and the pubic bone into two reciprocal golden rectangles. As far as their relations to each other are concerned, the length of both pubic bone and ischium tend to form with the width of the illium further golden, dinergic and reciprocal relations of neighbors.

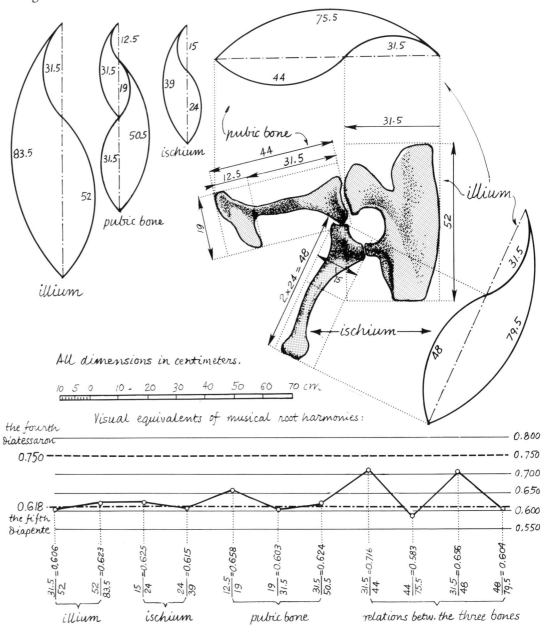

All dimensions in centimeters.

10 5 0   10 - 20   30   40   50   60   70 cm.

Visual equivalents of musical root harmonies:

the fourth
Diatessaron

0.750

0.618
the fifth
Diapente

$\frac{31.5}{52}=0.606$ $\frac{52}{83.5}=0.623$ $\frac{15}{24}=0.625$ $\frac{24}{39}=0.615$ $\frac{12.5}{19}=0.658$ $\frac{19}{31.5}=0.603$ $\frac{31.5}{50.5}=0.624$ $\frac{31.5}{44}=0.716$ $\frac{44}{75.5}=0.583$ $\frac{31.5}{48}=0.656$ $\frac{48}{79.5}=0.604$

0.800
0.750
0.700
0.650
0.600
0.550

illium   ischium   pubic bone   relations betw. the three bones

Fig. 116. Proportions of pelvic girdle of allosaurus.

From this extinct giant reptile, let us now turn to a small amphibian that is still with us: the frog. When the skeleton of a frog (fig. 117) is straightened out and the length of the bones are projected upon the central axis of the body, as shown in figure 118, the length of the bones show harmonically diminishing series as they move away from the pelvis. The top of the pelvis is the center of a series of circles which widen at a harmoniously diminishing rate and all joints appear to be located upon these circles.

The bar diagram at the right and the wave diagrams at the left show the harmonious nature of these series, giving a graphic image of the pulsations of growth, as they emanate from the pelvic center, very much like light vibrations radiating from the center of a lamp in the isocandle diagrams of figure 97.

The graph shows, in the familiar manner, how all shared proportional limitations approximate the three root harmonies of music. The central wave diagram gives the relationships between the extremities and the trunk, while the diagram at the left reveals how these very same proportional limitations also exist between the length of the skull, the spine, the pelvis, and the greatly extended jumping legs and feet.

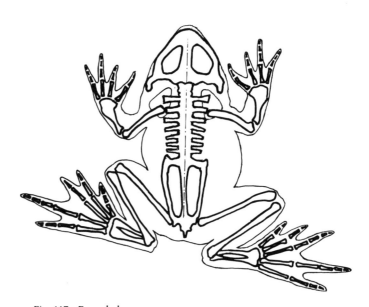

Fig. 117. Frog skeleton.

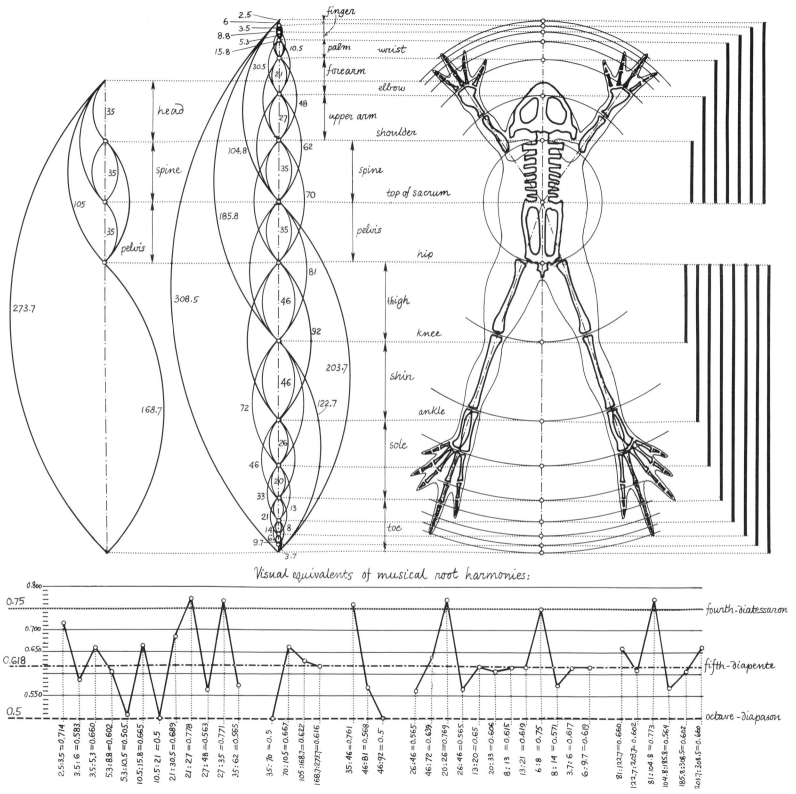

Fig. 118. Proportions of frog skeleton.

Skipping several rungs on the evolutionary ladder, we will now examine the skeleton of a horse. (fig. 119) Even in this reassembled state, the grace of its gallop is delightful. Figure 120 reveals some of the secrets of this grace—the proportional harmony that unites the diversities within the bone structure.

One aspect of this proportional relationship is between the skull and vertebral column on one side and the extremities on the other. (fig. 120) Lined up with their length-dimensions along their longitudinal axes, the parts of this anatomical structure show shared relationships between: skull and phalanges; neck and metatarsal-metacarpal bones; trunk and shinbone-forearm; and sacrum-tail and upper arm-thighbone.

The dimensioned wave diagrams further illustrate how these diverse articulations approximate musical root harmonies, as the graph demonstrates.

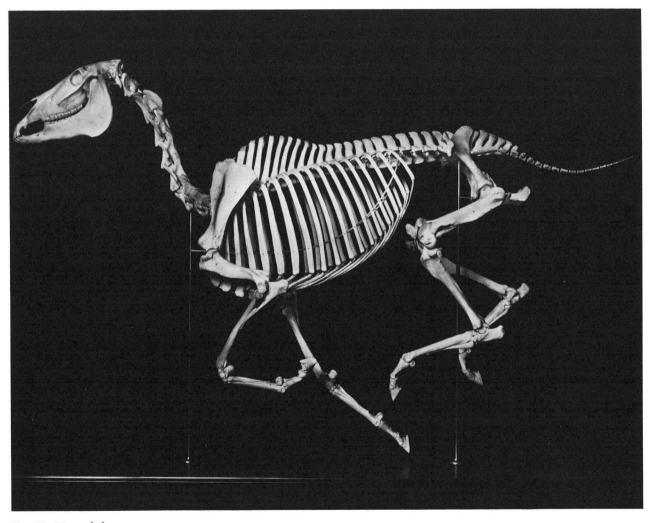

Fig. 119. Horse skeleton.

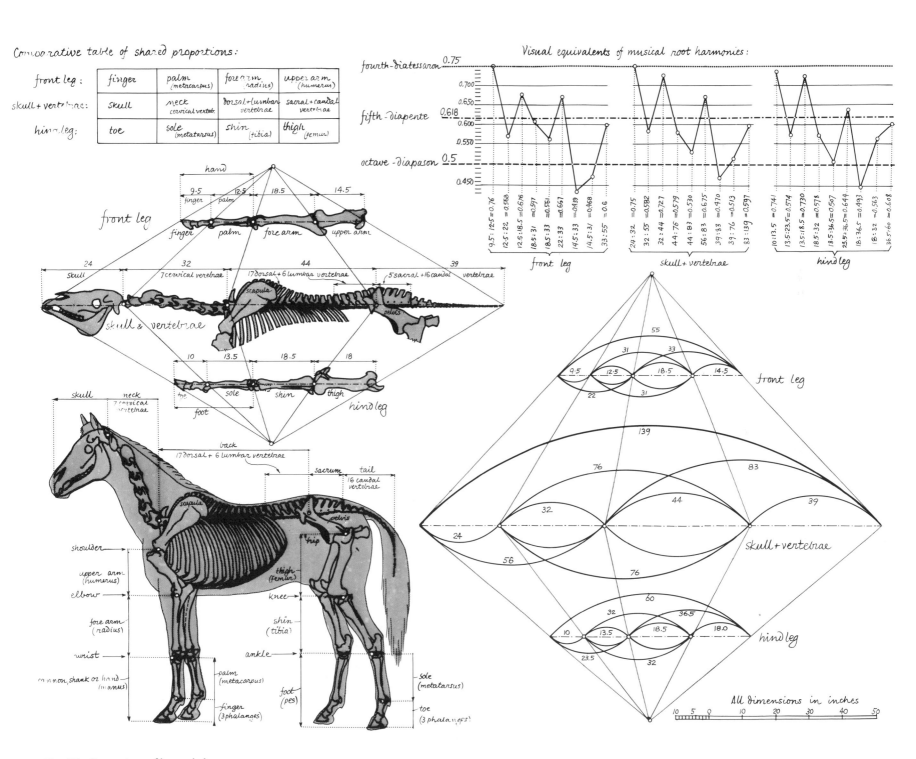

Fig. 120. Proportions of horse skeleton.

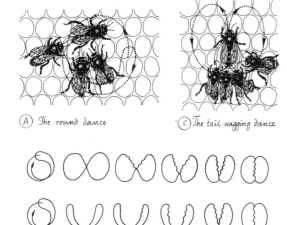

Fig. 121. Body language of a dog. Darwin's illustration of hostility, *above,* and devotion, *below.*

(A) The round dance   (C) The tail-wagging dance

(B) Transitions from round dance to tailwagging dance.

Fig. 122. Dance language of bees. **A.** Round dance; three bees are receiving message. **B.** Transition from round dance to tail-wagging dance. Schematic: *upper,* via figure eight form; *lower,* via sickle-shaped form. **C.** Tail wagging dance; four bees are receiving message.

We have seen several examples demonstrating the important role that shared proportional relationships play in the anatomy of animals. We shall end this chapter by attempting to show that such sharing is not limited to the proportional relationships of physical anatomy, but extends also to the social relationships of animals, thus supporting our view that sharing is indeed one of the basic pattern-forming processes of nature.

Sharing: the nature of nature

Sharing as a pattern-forming process is so basic in animal life that it is hard to select a few examples—suited to the limited scope of this presentation—without neglecting many more equally significant ones. Perhaps one could start with the sharing of parental care.

The Russian naturalist Peter Kropotkin, in his book *Mutual Aid* mentions that the lapwing (*Vanellus vanellus*) "has well merited the name 'good mother' which the Greeks gave it, for it never fails to protect other aquatic birds from the attacks of their enemies."[41] This brave bird is known for sharing the plight of its fellows by feigning a wound, thereby luring attackers away from a threatened nest. It accomplishes this by crying and flying with peculiarly flapping wings (thus the name "lapwing").

The lapwing's ingenious body language matches countless other examples of sharing by way of body language among animals. Darwin observed that dogs express their feelings of hostility or friendship through body movements, as illustrated in figure 121.

From the researches of von Frisch and others we know that bees share detailed communications about food sources by using the body language of dance. (fig. 122) A round dance is performed to convey that food is nearby (**A**). If the food is farther away, the dance is in a figure-eight pattern (**B**), and vigorous tail-wagging is added (**C**). Such dancing not only shares information: the dancer also shares energy and excitement, which induces fellow bees to swarm out and get the food, upon which their life depends.[42]

Mating dances of cranes and other birds are well known. But birds have also been reported to dance and sing in concert for no other apparent reason than shared delight, as W. H. Hudson observed in his book *Naturalist on the La Plata*, about his travels in South America, before the turn of the century. He remarked that he occasionally "saw a whole plain covered with an endless flock of Chakars (*Chauna chavarria*). . . . Suddenly the entire multitude of birds covering the marsh for miles burst forth in a tremendous evening song . . . it was a concert well worth riding a hundred miles to hear."[43]

The male cricket shares his mating urge with distant females whom he attracts with his mating call. Birds too sing to attract mates, and they also share all kinds of other information with each other, for instance warning other birds about approaching enemies, threatening intruders upon their territory, or calling fellow birds to gather. Even whales are known to communicate with each other by singing. The humpback whale (*Megaptera novaeangliae*) in particular has been reported to sing in eerie sounds beautiful even to the human ear. This song is said to be very loud and to cover a range from deep basso to high soprano. What messages whales share with each other by singing is not known, but scientists assume that it serves to "identify individuals, and to hold small groups together during the long transoceanic migrations."[44]

Fig. 123. Dolphins rescuing disabled fellow dolphin.

The migrations of animals are immense shared enterprises. Kropotkin observed the gathering of thousands of migratory birds on the steppes of Siberia, and said that "they come together at a given place for several days in succession before they start, and they evidently discuss the particulars of the journey. . . . All wait for the tardy . . . and finally they start in a certain well-chosen direction—a fruit of accumulated collective experience—the strongest flying at the head of the band and relieving each other in that difficult task. They cross the seas in large bands consisting of big and small birds."[45] One rarely appreciates the magnitude of these shared animal enterprises. The Arctic tern travels twice yearly from one polar area of the earth to the other, spending the summers at the North Pole along the shores of the Arctic Ocean and the winters around the South Pole in Antarctica, a travel distance of 11,000 miles each way. Such gigantic enterprises could not take place without superb cooperation and mutual assistance.

Sharing in each others' distress and rescuing endangered fellow animals is reported among many species. In 1969 it was observed in the Mediterranean how an electroharpooned dolphin was supported and eventually rescued by fellow dolphins. (fig. 123) Similar reports have been made about wild dogs, African elephants, and baboons, among others.[46] To quote Kropotkin once more about friendship and loyalty among animals: "When a parrot has been killed by a hunter, the others fly over the corpse of their comrade with shrieks of complaints and themselves fall victim of their friendship, as Audubon said."[47] Kropotkin was one of the first to

Fig. 124. Shared shapes of waves, hills, clouds and rocks. Pacific Ocean at Point Reyes, California, *left,* and rock formations at Glen Canyon, Colorado, *right.*

collect a substantial volume of scientific evidence of these basic patterns of animal behavior. He became convinced that sharing in the form of "mutual aid and cooperation" was not only the "prehuman origin" of all moral behavior but was also a basic condition of survival and a critical factor of evolution. [48]

In the nineteenth century this view was neglected in favor of popularized Darwinian concepts such as natural selection, survival of the fittest, struggle for existence, and competition. The controversy is still on between those who hold that human beings are inherently aggressive and competitive, and adherents of the opposite view that we have equally strong natural potentialities for nonviolent cooperation.

The "innate aggressiveness" theory has been revived and made popular by the writings of Konrad Lorenz, Robert Ardrey, Desmond Morris and William Golding, among others. On the other side are those who hold that war and violence are learned behaviors, culturally fostered. The anthropologist Ashley Montagu writes, "Human beings can learn virtually anything."[49] "War is not in our genes," says Sally Carrighar in *Man and Aggression.*[50] "While it is obvious that many men are killers, it is equally true that many more are not. . . . Man's propensity for violence is not a racial or a species attribute woven in his genetic fabric. It is culturally conditioned by history and the ways of life." So wrote Rene Dubos in 1971.[51]

To quote again Ashley Montagu, "At the present time the principle of cooperation is in a fair way to becoming established as an important factor in the survival of living groups. W. C. Allee sums up the modern point of view as follows: "The balance between the cooperative . . . and the egoistic is relatively close. Under many conditions the cooperative forces lose. In the long run, however, the group-centered, more altruistic drives are slightly stronger. . . . Our tendencies toward goodness, such as they are, are as innate as our tendencies towards intelligence; we could do well with more of both."[52] Another anthropologist, Colin M. Turnbull adds on the basis of his studies of non-literate, tribal cultures, ". . . if man has a seemingly limitless capacity for violence, for aggression, he has an equally great potential for nonviolence, and nonaggressivity."[53]

In terms of modern and practical economic policy—individual, national and international—the economist Barbara Ward comes to this conclusion: "When men or governments work intelligently and farsightedly for the good of others, they achieve their prosperity too . . . Generosity is the best policy . . . Our morals and our interests—seen in true perspective—do not pull apart."[54]

Sharing is creative. If we share what we have with our neighbors who have less, we gain more than if we had kept it all for ourselves. This is the paradox of sharing expressed by the words St. Paul attributed to Jesus: "It is more blessed to give than to receive."

We have seen that sharing as a basic pattern-forming process shapes harmonious relationships in animal and human life the same way it shapes proportional harmonies in animal anatomy, in music and in the other arts. There is indeed a "mana of sharing" throughout nature. If one looks at a view of the ocean shore, say in California, placing beside it a picture of rock formations, for instance in Colorado, (fig. 124) it is hard to tell where the rocks end and clouds or ocean begin, because the folds of the rocks and the clouds share the wavelines of water.

Similarly, a peacock's unfolding plumage shares the dinergic pattern at the center of a daisy. The dotted lines in figure 125 connecting the eyes of the peacock's plumage are identical with the logarithmic spirals used to reconstruct the daisy's pattern. When circles are inserted between the spiral lines (fig. 126), the peacock's pattern is transformed into the central pattern of the daisy. This is not magic: it is the mana of sharing, which is the very nature of nature.

Fig. 125. Eyes of peacock's tail are at intersecting points of logarithmic spirals.

Fig. 126. Circles show shared patterns of peacock and daisy.

CHAPTER 6: Order and Freedom in Nature

Organic and inorganic patterns

Sharing and dinergy are both basic pattern-forming processes that unite diversities. That there exists a basic unity within the manifold diversities of this world is one of the oldest observations of mankind. Ancient cultures attributed this unity to divinities, or to a single creator. The presocratic philosophers sought the secret of this unity in a universal substance—Thales saw it in water, Anaximenes in air, and Heraclitus in fire. This last philosopher is credited with having developed the concept of "unity in diversity."

In more recent times, this concept has become basic for both art and science. The American mathematician G. D. Birkhoff in 1928 developed a theory of aesthetic measure based on this principle, which he referred to as "order in complexity."[55] The measure of aesthetic value, according to this theory, is in direct proportion to order and in inverse proportion to complexity. H. E. Huntley, in his book about the golden section mentioned earlier, quotes J. Bronowski: "Science is nothing other than the search to discover unity in the wild variety of nature, or more exactly, in the variety of our experience. Poetry, painting, the arts are the same search, in Coleridge's phrase, for unity in variety."[56]

One of the most beautiful examples of this principle in nature is the snowflake: every one is different, yet all are united by their basic hexagonal pattern. (fig. 127) Each snowflake is restricted to one pattern, repeated and reflected twelve times (see triangles and arrows). Such uniformity is characteristic of all inorganic, crystalline patterns, which have more order and uniformity than living patterns.

Hexagonal patterns, such as the snowflake, are more common in inorganic than in organic nature, which favors pentagonal patterns. There are, however, interesting connections

Fig. 127. Snowflakes. Arrows indicate triangular pattern repeated 12 times in each flake.

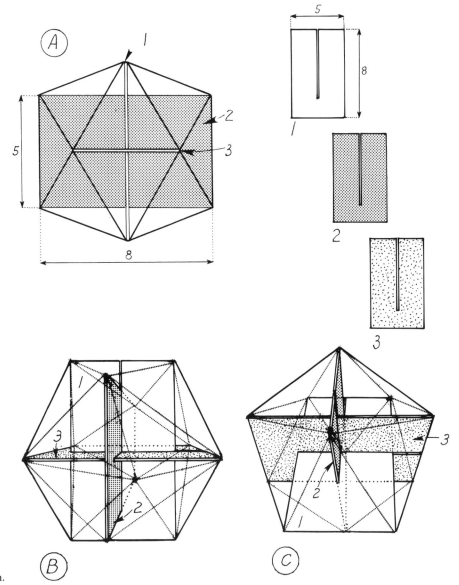

The icosahedron, which consists of 20 isosceles triangles can be constructed by making three equal golden rectangles out of cardboard, with a slit along the centerline (see at left) and then sliding them into each other at 90° eggcrate fashion. Twenty rubber bands fastened at slits in the corners of the golden rectangles complete the model, shown in orthogonal projection (A) where rectangles 1 & 3 appear edgewise. Perspective views (B) & (C) show the hexagonal and pentagonal outlines respectively.

Fig. 128. Model of icosahedron.

between hexagonal and pentagonal patterns. One of these appears in the icosahedron (fig. 128) consisting of twenty triangular faces, which has a hexagonal outline when viewed from one side, and a pentagonal one when viewed from the other side. Its diagonal planes (connecting diagonally opposite edges) are golden rectangles.[57] Perhaps the example of the icosahedron is a reflection of the relationship existing between inorganic and organic matter, the latter being built up from the former, as in the human body which is two-thirds water.

The unity within the diversities of organic and inorganic patterns is also seen in the spiral patterns of certain galaxies (fig. 129) which echo on a cosmic scale the minute dinergic spiral patterns of shells and flowers.

Fig. 129. Spiral galaxy's mandala pattern is similar to spirals of organic growth.

Pinecones likewise unfold in similar patterns. The seed-containing scales grow along the intersections of two sets of helixes—spirals that unfold in three-dimensional space like the DNA molecule. In plan-view, the spiral projections of these helixes resemble miniature galaxies. (fig. 130) In the cone of the Jeffrey pine, figure 131, thirteen helixes move in one direction and eight move in the other, closely approximating golden section proportions.

In the early 1900s in England, T. A. Cook published a profusely illustrated book, *The Curves of Life,* about the prevalence of the golden section's proportions in nature and art.[58] But he stressed not the paradoxical combination of unity and diversity, but rather diversity alone, as if unity would necessarily mean uniformity, which is fortunately not so. Two American books which appeared about the same time, *Nature's Harmonic Unity* and *Proportional Form* by Samuel Colman and C. Arthur Coan,[59] also describe the role of the golden section in the harmonious patterns of nature and art. But these books emphasize only the unity of these patterns. That diversity and unity are dinergically *joined* in all harmonies of nature and art is a point not stressed in these books, though it is the truth documented by all their examples.

Reality unites and diversifies at one and the same time. Montaigne observed in the fifteenth century: "As no event and no shape is entirely like another, so none is entirely different from

Fig. 130. Pattern of a pinecone.

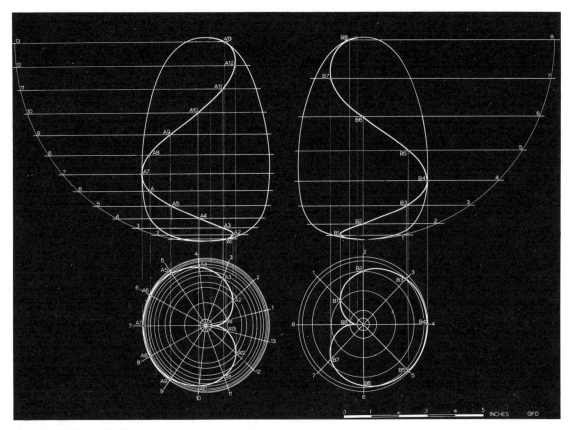

Fig. 131. Helixes of Jeffrey pinecone.

another . . . . If our faces were not similar, we could not distinguish man from beast; if they were not dissimilar we could not distinguish man from man."[60]

How nature accomplishes the seemingly impossible, creating forms that are both similar and dissimilar, united and diverse at the same time, has been brilliantly demonstrated by Sir D'Arcy Wentworth Thompson's "theory of transformations" to which earlier reference was made.[61] With the help of this theory, the shape of one species can be seen to derive from the shape of another, related species. This distinguished author derived, for instance, from the shape of the very common porcupine fish (*Diodon*), the shape of a very different-looking sunfish (*Orthagoriscus mola*) by transforming the right-angled network of coordinates, within which the former was plotted, into a corresponding but deformed network of curved lines that fitted the latter. (fig. 132) This could be done because, as the author says, "There is something, an essential and undisputable something, that is common to them."[62] What this essential something might be can perhaps be found by comparing the basic proportions of these two shapes: they are both variations of the golden section's proportions. The porcupine fish fits into two reciprocal golden rectangles, while the sunfish fits into two 3-4-5 triangles.

The zoologist Paul Weiss, in his inaugural address as president of the American Association

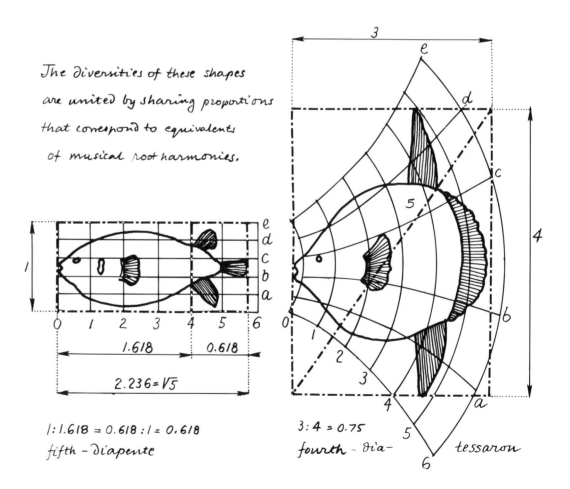

The diversities of these shapes are united by sharing proportions that correspond to equivalents of musical root harmonies.

Fig. 132. Comparative analysis of porcupine fish, *left,* and sunfish, *right.*

1.618

0.618

2.236 = √5

1 : 1.618 = 0.618 : 1 = 0.618
fifth - Diapente

3 : 4 = 0.75
fourth - Dia-   tessaron

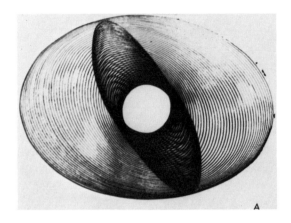

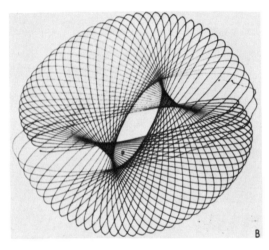

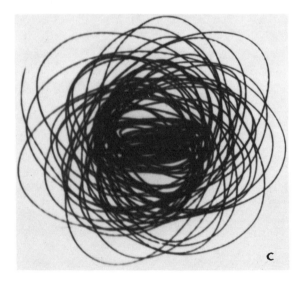

Fig. 133. Varieties of dinergic harmonograph patterns.

for the Advancement of Science in 1953, interpreted the unity within diversities in the patterns of organic nature as a combination of order and freedom. This combination is as paradoxical as it is dinergic: order and unity involves constraint, while diversity implies the freedom to differ. "Life is order, but order with tolerances."[63]

Snowflakes have shown us that this principle is also present in inorganic nature. Other inorganic patterns that demonstrate the principle of order and freedom can be seen in the movement of pendulums. The shapes in figure 133 were made by a harmonograph, which consisted of two pendulums that registered their simultaneous movements with the help of a tablet attached to one and a pen attached to the other. When their lengths—which determine the swing times of the two pendulums—were left entirely free of any constraints or relationship to each other, the pattern became chaotic, like tangled wool. (**C**) When, however, the lengths were adjusted so that their swing times related to each other in ratios expressible in small whole numbers, then the patterns became harmonious, as **A, B,** and **D** illustrate. The pendulums still swung freely, but now the lines also reflect, by the number of looped configurations on each side, the proportional limits between the length, and thus between the swing times, of the generating pendulums.[64] The difference between chaotic and harmonious lines produced here by gravity is analogous to the difference between noise and musical sound discussed earlier.

There are curious similarities between harmonograph patterns and certain shell shapes. Figure 134 shows the cross section of a nautilus shell and a harmonograph pattern produced by two pendulums that were equally long but swinging in opposite directions. The two patterns are diverse in origin, but there is a similarity between their graceful lines. This similarity is rooted in the fact that dinergic proportional relationships create both patterns: increments of growth in the shell and swing times in the pendulums.

Gravity is weight. The lifting of weight by its dinergic sharing is grace. It is the seemingly effortless release from burden that charms us in grace, hence its divine connotation. *Gravity and Grace* is the title of Simone Weil's posthumously edited notebooks in which she movingly records her observations about the strange ways in which necessity and beauty, order and freedom, gravity and grace, unity and diversity are linked together in nature as well as in human destiny.[65]

To our previous examples of unity in diversity, let us now add a few from the worlds of insects and of humans.

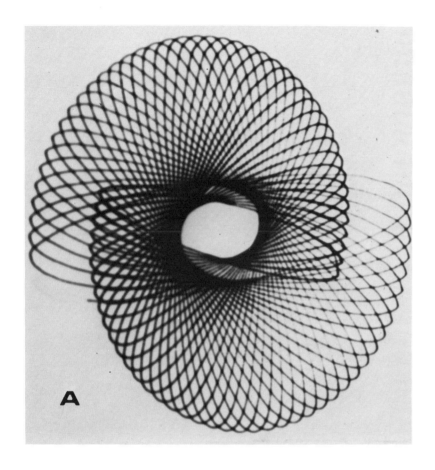

Fig. 134. **A.** Pattern drawn by harmonograph.
B. Cross section of chambered nautilus shell.

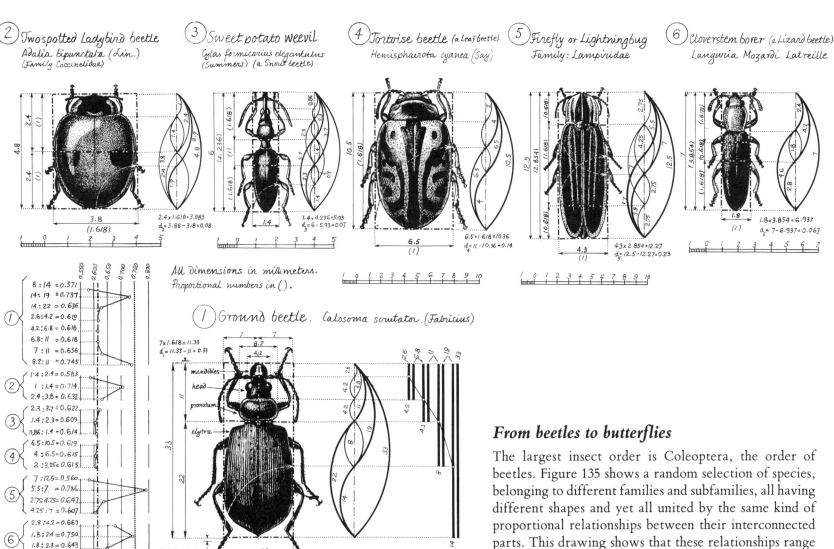

Fig. 135. Unity in the diversities of beetle shapes.

From beetles to butterflies

The largest insect order is Coleoptera, the order of beetles. Figure 135 shows a random selection of species, belonging to different families and subfamilies, all having different shapes and yet all united by the same kind of proportional relationships between their interconnected parts. This drawing shows that these relationships range between the 0.618 ratio of the golden section and the 0.75 ratio of the Pythagorean triangle. The graph at the lower left shows the diverse ways in which these proportional relationships are approximated. As a result of these shared relationships, the overall proportions as well as the detailed articulations of every one of these beetle shapes approximate the visual equivalents of musical root harmonies of fifth-diapente and fourth-diatessaron.

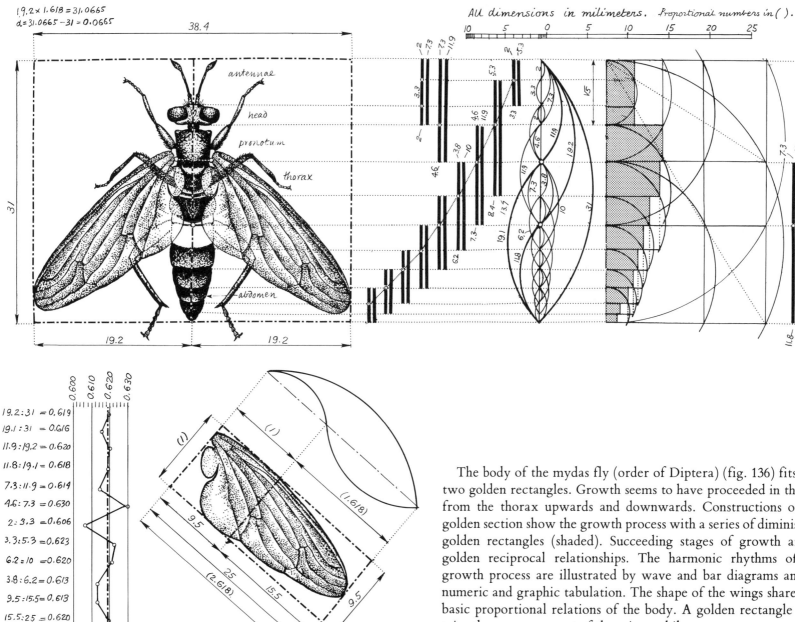

Fig. 136. Proportional analysis of mydas fly.

The body of the mydas fly (order of Diptera) (fig. 136) fits into two golden rectangles. Growth seems to have proceeded in this fly from the thorax upwards and downwards. Constructions of the golden section show the growth process with a series of diminishing golden rectangles (shaded). Succeeding stages of growth are in golden reciprocal relationships. The harmonic rhythms of the growth process are illustrated by wave and bar diagrams and by numeric and graphic tabulation. The shape of the wings shares the basic proportional relations of the body. A golden rectangle contains the narrower part of the wing, while a square corresponding to the width of the wing fits the broader part, the relation between square and rectangle being the reciprocal "relation of neighbors."

Though dragonflies, damselfies, darners, and skimmers (order of Odonata) don't look like flies, they nevertheless share the same basic proportional limitations. The common skimmer's overall shape, for instance, neatly fits into a single golden rectangle. (fig.

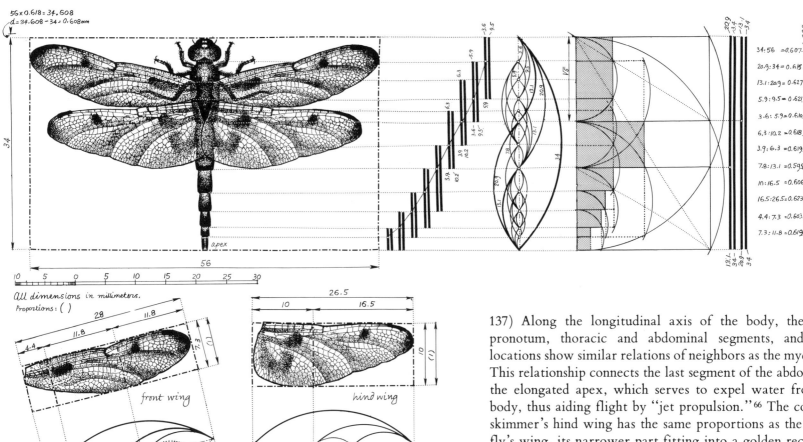

All dimensions in millimeters.
Proportions: ( )

front wing

28 : 3.854 = 7.265
d = 7.3 − 7.265 = 0.035 mm

hind wing

26.5 : 2.618 = 10.12
d = 10.12 − 10 = 0.12 mm

Fig. 137. Proportional analysis of common skimmer.

137) Along the longitudinal axis of the body, the head, pronotum, thoracic and abdominal segments, and wing locations show similar relations of neighbors as the mydas fly. This relationship connects the last segment of the abdomen to the elongated apex, which serves to expel water from the body, thus aiding flight by "jet propulsion."[66] The common skimmer's hind wing has the same proportions as the mydas fly's wing, its narrower part fitting into a golden rectangle, while the broader portion is contained in a square equal to the wing width. The elongated front wing fits into two golden rectangles and their reciprocal.

The proportions of the wings of other insects, such as mayflies, bees, wasps, ants, crickets, and grasshoppers, show similar proportional limitations. What is more, the same wing shapes appear outside the animal kingdom, for instance in the winged seeds of the maple tree. Figure 138 shows four random pairs of winged maple seeds. In specimen **1,** the two wings completely overlap, fitting into a golden rectangle and its reciprocal. In specimen **2,** the two wings fit into a single golden rectangle, and the widest point of the two combined wings is at the golden section point of its length. In specimen **3,** the right wing has grown along the diagonal of a golden rectangle, while the left wing fits into a golden rectangle and a square, the width of which corresponds to the full width of the wing (in the position it occupies within the whole configuration). Specimen **4** shows a further variation, each of the wings corresponding to the diagonals of 3:4 rectangles.

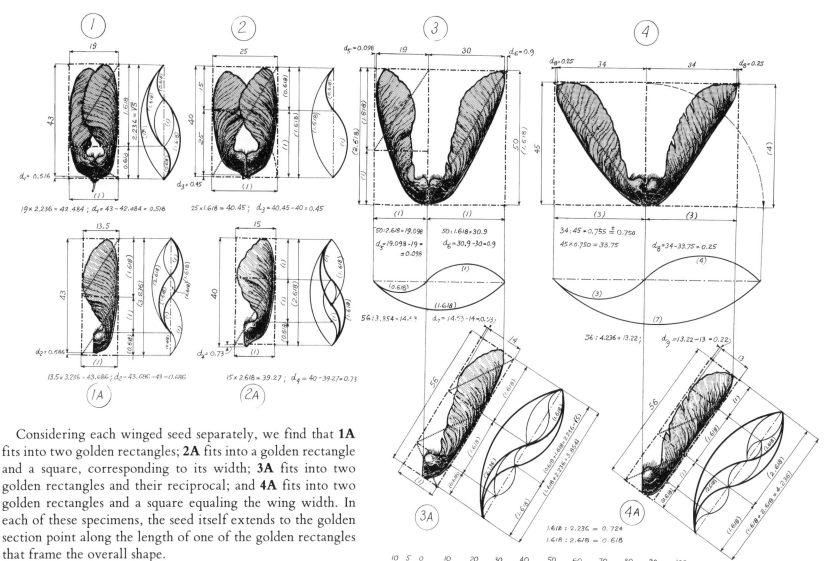

Considering each winged seed separately, we find that **1A** fits into two golden rectangles; **2A** fits into a golden rectangle and a square, corresponding to its width; **3A** fits into two golden rectangles and their reciprocal; and **4A** fits into two golden rectangles and a square equaling the wing width. In each of these specimens, the seed itself extends to the golden section point along the length of one of the golden rectangles that frame the overall shape.

The variety and beauty of butterflies (order of Lepidoptera) is proverbial. The proportioning of butterfly wings varies from species to species, yet there is a unity in these diversities, a unity created by the same proportional limitations we observed earlier, with the 1:2 = 0.5 proportion added, corresponding to the octave-diapason.

For each butterfly illustrated here, six basic proportional relationships were checked: total or half the wing width (**2A** or **A**) to wing height (**B**); body length (**C**) to total wing height (**B**); body length (**C**) to total or half the wing width (**2A** or **A**); length of hind wing (**E**) to length of front wing (**D**); width of front wing (**G**) to its own length (**D**); and width of hind wing (**F**) to its own length (**E**).

All dimensions in millimeters. Proportional numbers in ( ).

Fig. 138. Proportional patterns of winged maple seeds.

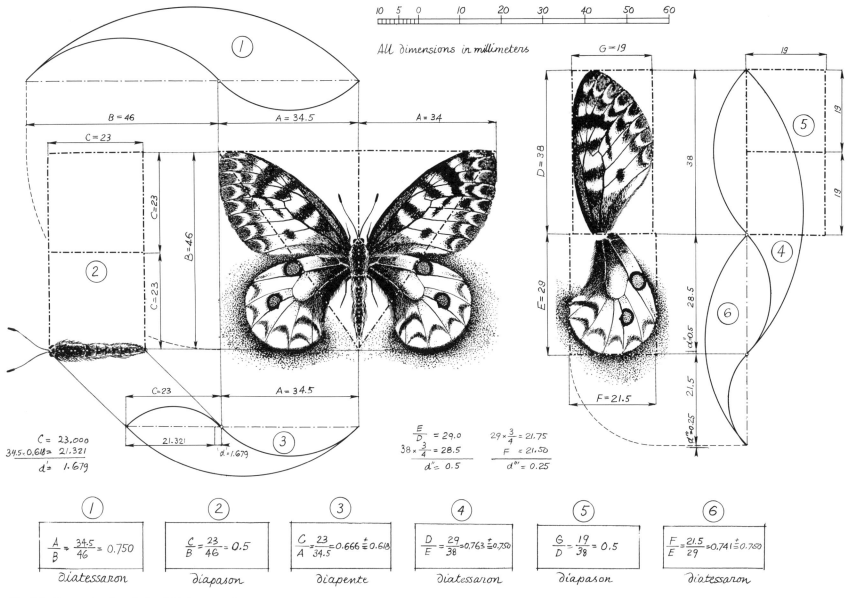

10 5 0 10 20 30 40 50 60

All dimensions in millimeters

$B = 46$ $A = 34.5$ $A = 34$

$C = 23$

$C=23$ $B=46$ $C=23$

$C=23$ $A = 34.5$

$G = 19$ 19

$D = 38$ 38   19   19

$E = 29$ 28.5

$F = 21.5$

$d' = 0.5$ 21.5 $d'' = 0.25$

21.321 $d' = 1.679$

$C = 23.000$
$34.5 \times 0.618 = 21.321$
$d' = 1.679$

$$\frac{E}{D} = 29.0 \qquad 29 \times \frac{3}{4} = 21.75$$
$$38 \times \frac{3}{4} = 28.5 \qquad F = 21.50$$
$$d'' = 0.5 \qquad d''' = 0.25$$

①	②	③	④	⑤	⑥
$\dfrac{A}{B} = \dfrac{34.5}{46} = 0.750$	$\dfrac{C}{B} = \dfrac{23}{46} = 0.5$	$\dfrac{C}{A} = \dfrac{23}{34.5} = 0.666 \lessgtr 0.618$	$\dfrac{D}{E} = \dfrac{29}{38} = 0.763 \lessgtr 0.750$	$\dfrac{G}{D} = \dfrac{19}{38} = 0.5$	$\dfrac{F}{E} = \dfrac{21.5}{29} = 0.741 \lessgtr 0.750$
diatessaron	*diapason*	*diapente*	*diatessaron*	*diapason*	*diatessaron*

Fig. 139. Proportional analysis of *Clodius parnassius*.

Figure 139 shows a specimen of *Clodius parnassius,* belonging to the family of Parnassadiae: a light gray butterfly with distinctive round, red dots on its hind wings and crescent-shaped dark markings around the edges of the wings. It shows octave-diapason relationships between body length and total wing height (**2**), and between width and length of the front wings (**5**). Fifth-diapente harmony is approximated by the relationship of body length to half wing width (**3**), while fourth-diatessaron appears in three relationships: half the wing width to wing height (**1**), length of hind wing to length of front wing (**4**), and width to length of hind wing (**6**). (Differences between mathematically exact and actual proportions are indicated with **d** in the drawing.)

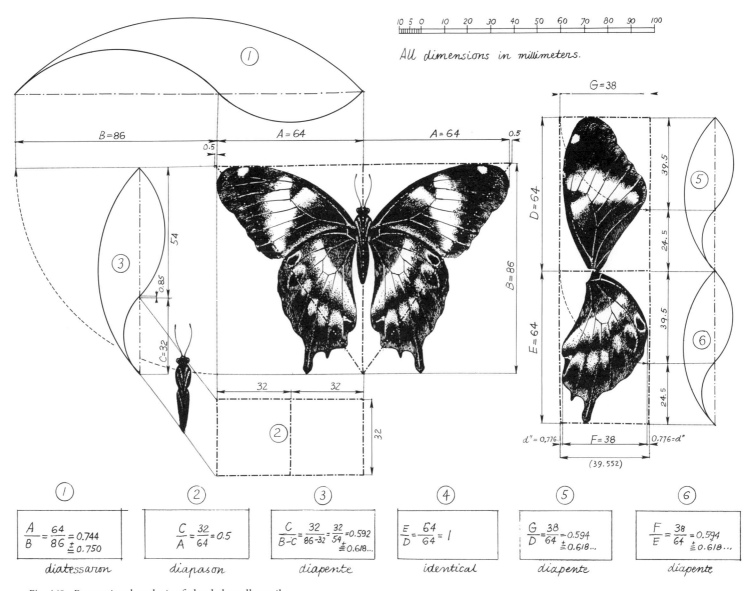

All dimensions in millimeters.

①	②	③	④	⑤	⑥
$\dfrac{A}{B}=\dfrac{64}{86}=0.744$ $\pm\ 0.750$	$\dfrac{C}{A}=\dfrac{32}{64}=0.5$	$\dfrac{C}{B-C}=\dfrac{32}{86-32}=\dfrac{32}{54}=0.592$ $\cong 0.618...$	$\dfrac{E}{D}=\dfrac{64}{64}=1$	$\dfrac{G}{D}=\dfrac{38}{64}=0.594$ $\cong 0.618...$	$\dfrac{F}{E}=\dfrac{38}{64}=0.594$ $\cong 0.618...$
diatessaron	*diapason*	*diapente*	*identical*	*diapente*	*diapente*

Fig. 140. Proportional analysis of clouded swallowtail.

Figure 140 represents a clouded swallowtail in ventral view. It is a large, beautiful butterfly, belonging to the family of Papilionidae, with wings that appear woven of dark lace, and with the characteristic elongation of the hind wing which accounts for its name and makes the hind wing as long as the front wing. We find the equivalent of diapason in the relationship of body length to half the wing width (**2**), while diatessaron relates half the wing width to the wing height (**1**). There are three diapente relationships in this specimen: one is between body length and wing height (**3**), the other two are between the width and length of each of the wings, (**5** and **6**).

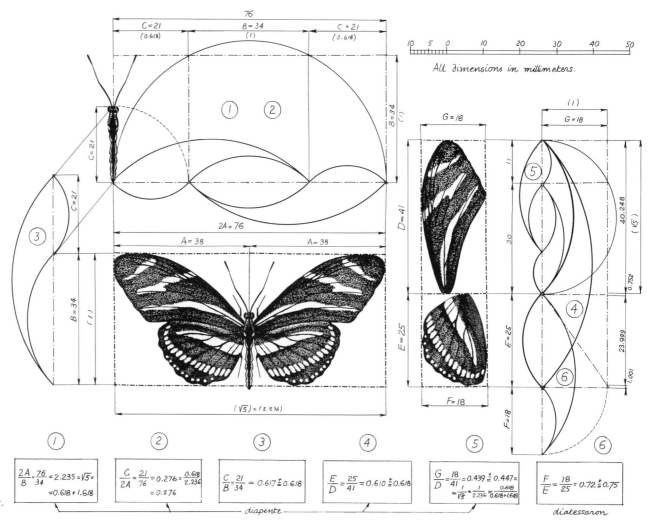

All dimensions in millimeters.

$$\frac{2A}{B} = \frac{76}{34} = 2.235 = \sqrt{5} = \\ = 0.618 + 1.618$$

$$\frac{C}{2A} = \frac{21}{76} = 0.276 = \frac{0.618}{2.236} \\ = 0.276$$

$$\frac{C}{B} = \frac{21}{34} = 0.617 \stackrel{+}{=} 0.618$$

$$\frac{E}{D} = \frac{25}{41} = 0.610 \stackrel{+}{=} 0.618$$

$$\frac{G}{D} = \frac{18}{41} = 0.439 \stackrel{+}{=} 0.447 = \\ = \frac{1}{\sqrt{5}} = \frac{1}{2.236} = \frac{0.618}{0.618 + 1.618}$$

$$\frac{F}{E} = \frac{18}{25} = 0.72 \stackrel{+}{=} 0.75$$

diapente ———— diatessaron

Fig. 141. Proportions of zebra butterfly.

The zebra butterfly, (fig. 141) (*Heliconius charitonius*), which belongs to the family of Heliconidae, is dark brown with bright yellow stripes. It has elongated front wings which fit into the $\sqrt{5}$ rectangle, composed of two reciprocal golden rectangles. An enlarged version of this same proportion encompasss the overall shape of this butterfly. The 21-millimeter body length and the 34-millimeter total wing width of this specimen correspond to Fibonacci numbers, bringing about a multiplicity of diverse diapente harmonies (see 1, 2, 3, 4, 5). The proportions of the hind wing approximate 3:4 = 0.75, the visual equivalent of diatessaron.

It is amazing to behold such unity in the manifold diversities of nature, each species developing freely its uniqueness, yet all united by sharing the same simple, dinergic and harmonious proportional limitations. Butterflies flutter, dragonflies dart, and beetles bore. The allosaurus must have moved with a heavy, lumbering gait, unlike the graceful gallop of the horse, the slow crawl of the crab, and the sudden leap of the frog. How unique each is, and how different than winged maple seeds, daisies and sunflowers, and yet they all are united by sharing the proportional limitations of musical root harmonies. This unity of diversities extends also to the proportions of human anatomy, as the following examples intend to show.

Human harmonies

The perception of human proportions has varied greatly throughout the ages. One of the earliest written documents dealing with human proportions is by Marcus Vitruvius Pollio, the first-century Roman architect and writer. He begins his *Ten Books on Architecture* with the recommendation that temples, in order to be magnificent, should be constructed on the analogy of the well-shaped human body, in which, he says, there is a perfect harmony between all parts.[67] We shall see more of Vitruvius's ideas on temples in the next chapter. Here we are concerned with his examples of harmonious human proportions. Among these he mentions that the height of a well-shaped man is the same as the span of his outstretched arms, these two equal measures yielding a square which encompasses the whole body, while the hands and feet touch a circle centered upon the navel.

This relatedness of the human body to the circle and the square rests upon the archetypal idea of "squaring the circle," which fascinated the ancients, because these shapes were considered perfect and even sacred, the circle having been looked upon as a symbol of the heavenly orbits, the square as a representation of the "foursquare" firmness of the earth. The two combined in the human body suggests in the language of symbolic patterns that we unite within ourselves the diversities of heaven and earth, an idea shared by many mythologies and religions.

When the Renaissance rediscovered the classical remains of Greece and Rome, Leonardo da Vinci illustrated Vitruvius's version of this idea with his own famous drawing (fig. 142). The bar diagrams and the triangular diagram, which here are added to the drawing, show how the adjacent parts of this body share proportions which fall within the range of the golden section

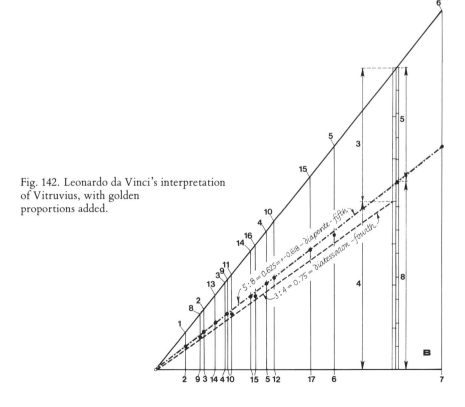

Fig. 142. Leonardo da Vinci's interpretation of Vitruvius, with golden proportions added.

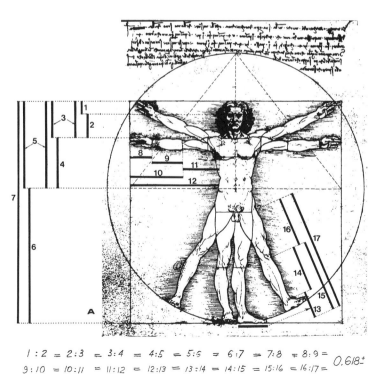

$$1:2 = 2:3 = 3:4 = 4:5 = 5:6 = 6:7 = 7:8 = 8:9 =$$
$$9:10 = 10:11 = 11:12 = 12:13 = 13:14 = 14:15 = 15:16 = 16:17 = 0.618^+$$

Fig. 143. Fra Luca Pacioli, author of *Divina Proportione,* with a student.

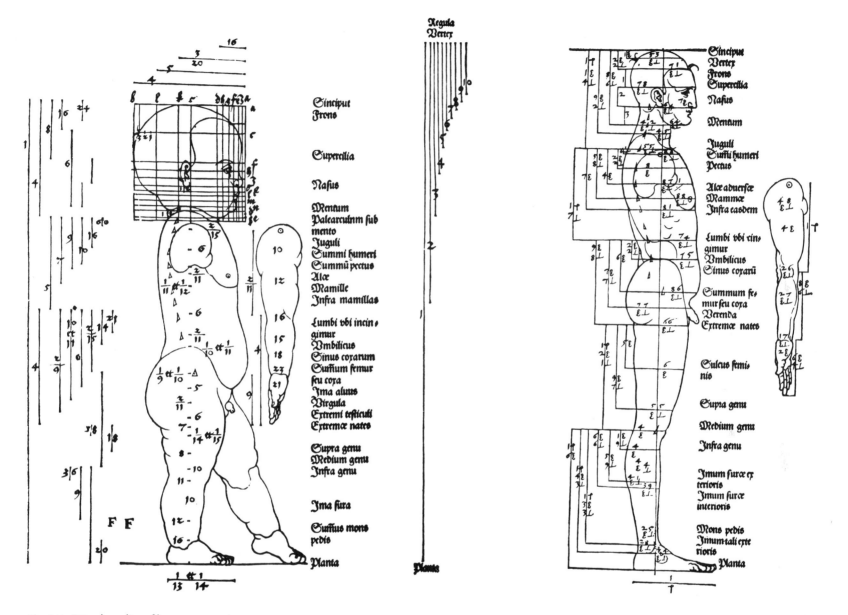

Fig. 144. Dürer's studies of human proportions using harmonic scales.

and the Pythagorean triangle. Leonardo, along with many of the Renaissance masters, was a great student of harmonious proportions and illustrated the mathematician Luca Pacioli's book *Divina Proportione,* about the golden section, published in 1509. (fig. 143) Leonardo summarized his studies of good proportions in memorable words: ". . . every part is disposed to unite with the whole, that it may thereby escape from its incompleteness."

This disposition of the human body's diverse parts to unite with the whole also fascinated another great Renaissance painter, Albrecht Dürer, who published several volumes about human proportions. His theories include the use of harmonic scales, as shown in figure 144, illustrating these relationships along drawings of a child's and a man's bodies.

The idea that the root harmonies of music—in accordance with revived Pythagorean concepts—correspond to good proportions in the human body, and should therefore also be followed in architecture, became a leading idea among the masters of the Renaissance. Following that time, some of these ideas about human harmonies took a mystical turn, strengthened by studies of Kabbalah, the Jewish mystical tradition, which became available in Latin translations of ancient Hebrew texts. The Englishman Robert Fludd depicted man as a microcosm, united with the macrocosm of the universe, combining dark, earthly potentialities with light, heavenly ones, attuned to universal musical harmonies like a monochord stretching from earth to heaven.[68] (fig. 145)

Later, the Ages of Enlightenment and Rationalism frowned upon such mystical ideas. The painter Hogarth considered it "a strange notion" that there should be any correspondence between beauty as seen by the eye and harmony as heard by the ear. The Scottish philosopher Hume made the point that beauty is in the eye of the beholder and that it is entirely subjective. The Englishman Edmund Burke said that "no two things can have less resemblance or analogy than a man, a house or a temple." By the end of the nineteenth century, John Ruskin asserted that "proportions are infinite as possible airs in music, and it must be left to the inspiration of the artist to invent beautiful proportions."[69]

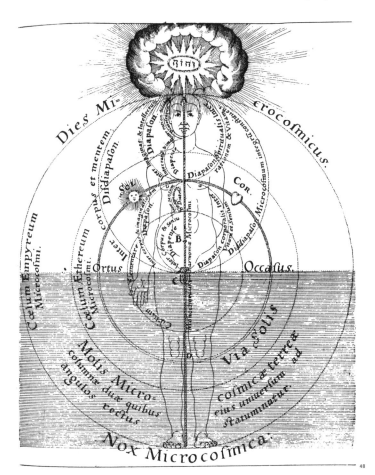

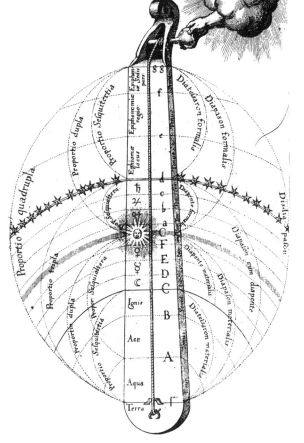

Fig. 145. Illustrations from Robert Fludd's *Utriusque Cosmi Historia.*

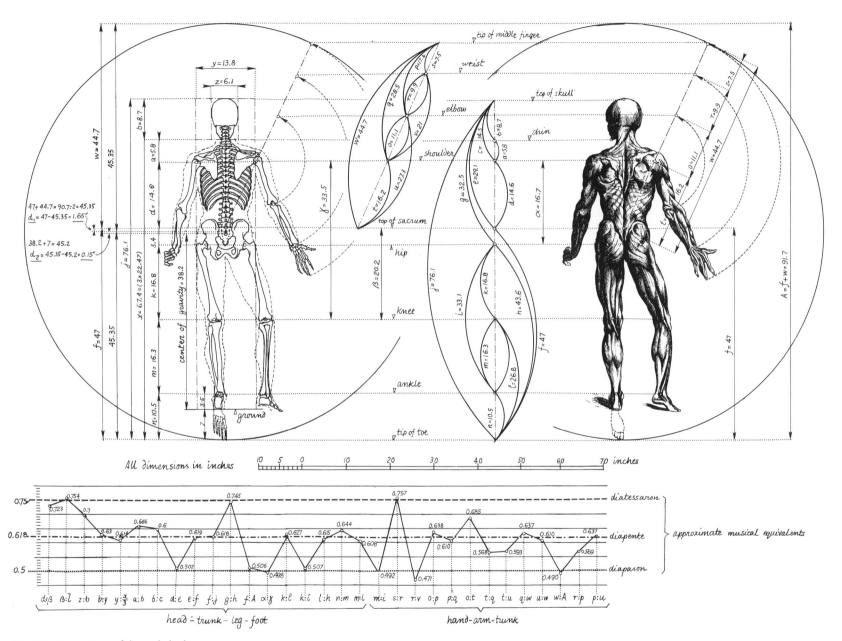

Fig. 146. Proportions of the male body.

Figures 146 and 147 are based upon the author's measurements of human skeletons, integrated with information gleaned from measurements of live models as well as from anatomical textbooks. All dimensions are those of average-sized adults as reported by U.S. governmental surveys.

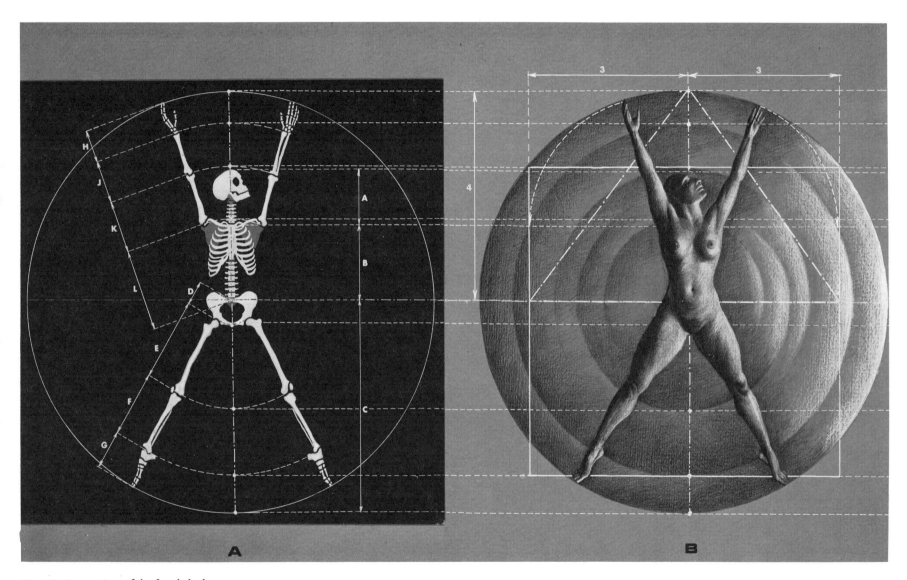

Fig. 147. Proportions of the female body.

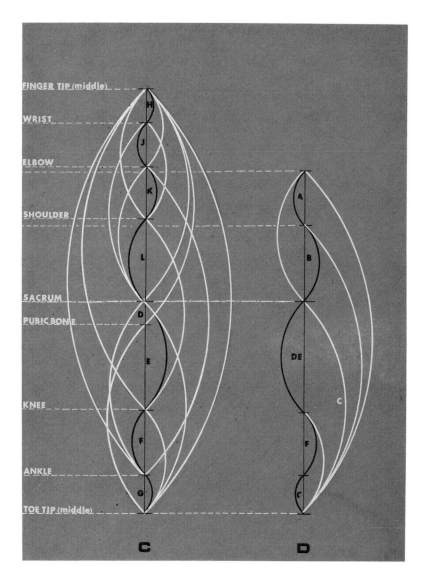

These drawings of the female figure show that raised arms and outstretched legs touch yet another circle, beyond the one centered at the navel. The center of this new circle is close to the center of gravity of the whole body, near the point where the vertebral column emerges from the sacrum, as in the frog. This circle is somewhat larger than Vitruvius's and Leonardo's, because they both refer to a figure standing on the soles of the feet, thus excluding the length of the feet themselves, while the present drawings include the length of the feet, fully extended like a ballet dancer's.

The diagrams in figures 146 and 147 show again how all parts of the human body share the same proportional limitations. Thus, the length relationships of hand to arm to trunk (to the starting point of the spine in the pelvis) are shared, as in the thoroughbred horse, by the relationships of head to neck, trunk, legs, and feet. The entire human bone structure fits neatly into three golden rectangles and a reciprocal, the latter containing the head. (See Appendices I & II.)

The unity we share with plants and animals is again visible from the fact that our growth, like theirs, seems to unfold from a single center, which in our case is at the top of the sacrum, just as in the frog. It may be recalled how the spirals of the daisy and sunflower also unfold from the center.

Because of its central location within the animal body, the ancients considered the sacrum of their sacrificial animals particularly sacred, hence its name: *os sacrum,* "sacred bone" in Latin. It is the center of the circle around the extended extremities, as shown in the skeleton in figures 146 & 147, and it is also the center of gravity of the entire body.

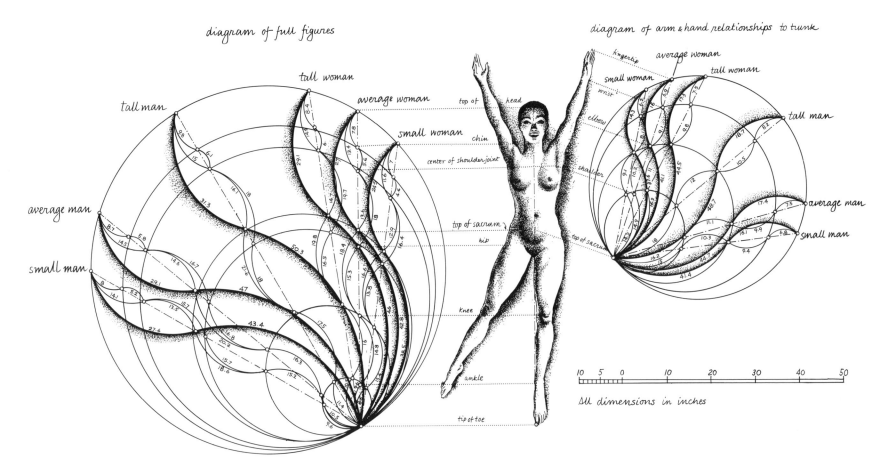

diagram of full figures

diagram of arm & hand relationships to trunk

Fig. 148. Unity of longitudinal proportions in women and men of diverse sizes.

A further instance of unity in diversity within human anatomy is the surprising correspondence between proportions of diversely sized bone structures of both sexes, as the comparative table in Appendix I indicates. Figure 148 shows that all corresponding longitudinal dimensions of tall, average and small women *and* men tend to fall upon coaxial circles.

The wave diagrams of figure 148 are simplified versions of the ones seen earlier, with proportions corresponding to the ones in Appendices I and II. The trend towards unity of these wave patterns is so marked that they are almost identical, appearing to be stroboscopic images of one and the same moving pattern. These rhythmic waves are reminiscent of the gracefully swinging loops of pendulum patterns and the fanlike overlapping stages of growth in the daisy's central pattern.

The graph in Appendix I indicates the spread of approximations to musical root harmonies. This graph could be looked upon as a kind of musical score of the human body's silent harmonies, or rather as variations upon this singular theme song.

As great as the diversities of female and male bodies are, still women and men are united by the almost complete identity of their anatomic proportions, at least as far as the length of their bones are concerned. The only difference is a general refinement in the measurements of the female skeleton and a widening of the pelvic girdle.

Finally, we find an astonishing unity between the proportional harmonies of the whole body and its diverse parts. Figure 149 shows a man's hand, traced from an x-ray photo. Even though this hand is somewhat deformed by arthritis, the length of the bones lined up on their axes show that all joints tend to fall upon coaxial circles. The rendering of the wave diagrams emphasizes the united rhythm of these proportions. The hand is a microcosm mirroring the macrocosm of the body. It grows out of the wrist as the spine grows out of the sacrum, and as wings grow out of the butterfly, or as leaves and flowers grow out of their stems.

Fig. 149. Unity and diversity of proportions in the human hand.

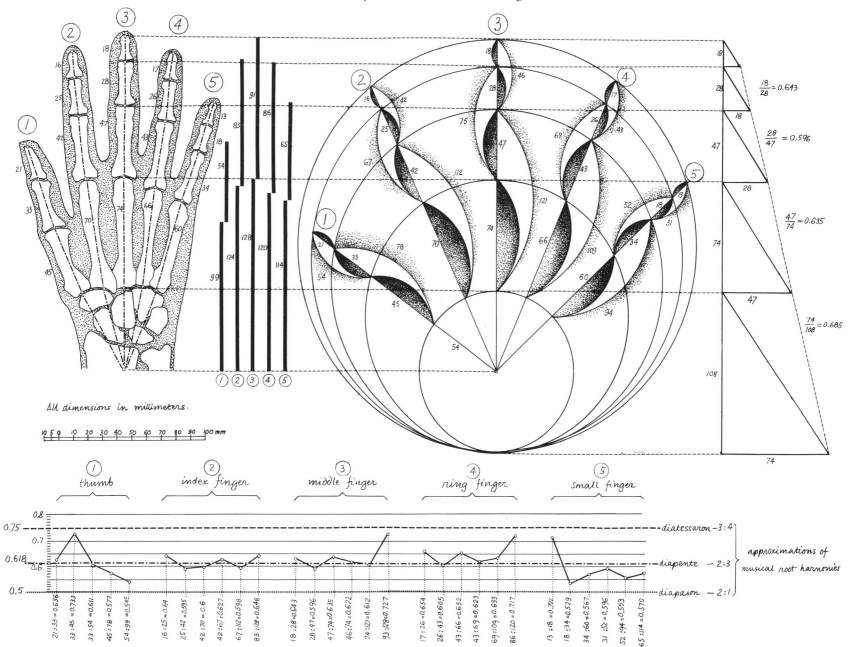

"Beauty is the harmony and concord of all parts, achieved in such a manner that nothing could be added or taken away or altered except for the worse."[70] These are the words of another master of the Renaissance, Leon Battista Alberti, an architect and author of a famous treatise on architecture.

The human body has the potentials for such harmony and beauty. This is obvious in the graceful movements of an accomplished ballet dancer, for instance. (fig. 150) The *American Heritage Dictionary* gives as the first meaning of the word "grace": "seemingly effortless beauty or charm of movement, form or proportion." Ballet is certainly anything but effortless, like any other form of art, but there is a freedom of movement in ballet that appears miraculously easy. This freedom turns upon order, the order of discipline that prepares and sustains all great accomplishments.

One of the secrets of this order in ballet is the support of the body's entire weight upon a single point: the center of gravity in the sacrum, which is upheld by only one outstretched leg on the tip of the toes. Growth, gravity and grace concentrated in one point of our sacrum is the power of limits enshrined in our bones.

Simone Weil, whose thoughts about grace and gravity were mentioned earlier, notes in another volume of her notebooks: "Something infinitely small, under certain conditions, operates in a decisive manner. There is no mass so heavy but that a given point is equal to it; for a mass will not fall if a single point in it is upheld, provided that this point be the center of gravity." She added—and this was during the days of Nazi occupation in France—"brute force is not sovereign in this world. . . . What is sovereign . . . is limit. . . . Every visible and palpable force is subject to an invisible limit, which it shall never cross. In the sea a wave mounts higher and higher, but at a certain point . . . it is arrested and forced to descend. . . . That is the truth which bites at our hearts every time we are penetrated by the beauty of the world. That is the truth which bursts forth in matchless accents of joy in the beautiful and pure parts of the Old Testament, in Greece among the Pythagoreans and all the sages, in China, with Lao Tzu, in the Hindu scriptures, in Egyptian remains."[71]

The truth of these words has been one of the main inspirations of this book. These words reveal the other, traditional meaning of the word grace: mercy and the divine gift of love, springing from the relatedness of all that exists.

The ballet dancer's visible grace is a symbol of the other, invisible grace: the potential of harmony and beauty that exists within every human being, for indeed all of us, not only accomplished dancers, have an inner center, a sacrum, both in a physical and a spiritual sense. That there also exist disorder and disease is testimony to nature's tolerance of diversity and freedom, which at times can be very generous. Alberti's statement holds true for the art of living as well as for other arts: too much added can harm as surely as too much taken away. Too much leisure and luxury can destroy our native beauty as much as too little of life's necessities. Harmony and grace are born from the marriage of plenty and poverty as Plato told us in his *Symposium*. A comparative study of classical Western and Far Eastern arts strengthens this view.

Fig. 150. Deborah Hadley, solo dancer of Pacific Northwest Ballet.

CHAPTER 7: Hellas and Haiku

Man the measure

The art forms of the East and West are vastly different: compare a Greek Apollo and a Tibetan Buddha, the Parthenon and a pagoda, or the epic poetry of Virgil and the haiku poetry of Japan. But the points where the arts of East and West do meet reveal an underlying human unity beyond the surface diversities.

"Man is the measure of all things," according to Protagoras, the Greek philosopher of the fifth century B.C. This epigram becomes palpable when one looks at great Greek sculpture.

The spearbearer or Doryphoros (fig. 151) is also from the fifth century B.C., created by Polycleitos, who is credited with having written a celebrated treatise, now lost, about the

Fig. 151. Doryphoros, the spear bearer.

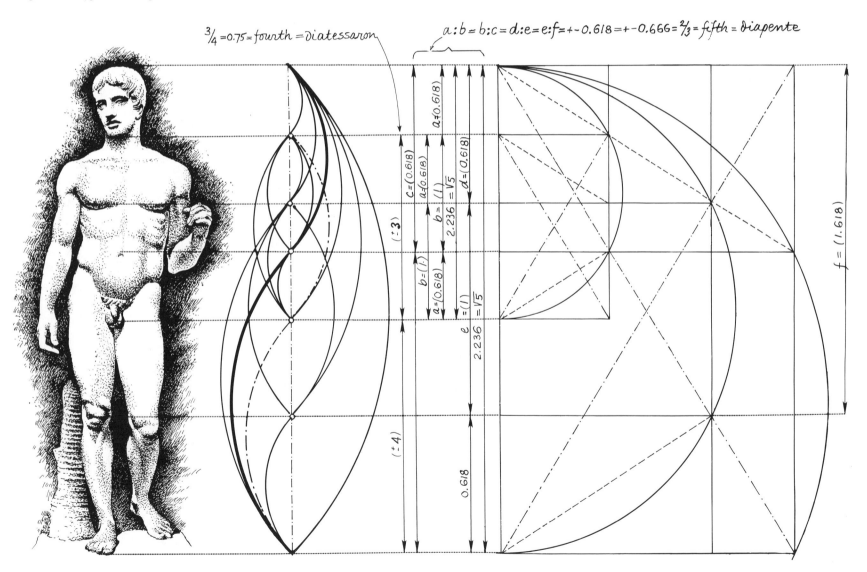

proportions of the human body. Golden section constructions of this sculpture show two sets of reciprocal golden rectangles, each set $\sqrt{5}$ long: the larger set encompasses the whole body, with knees and breasts at golden section points; the smaller set reaches from the top of the head to the genitals. The navel is at the golden section point of the full height, the genitals at the 3/4 point of the height to the chin. In the Aphrodite of Cyrene (fig. 152), one can recognize similarly harmonious length relationships, though the head is unfortunately missing.

Fig. 152. Aphrodite of Cyrene.

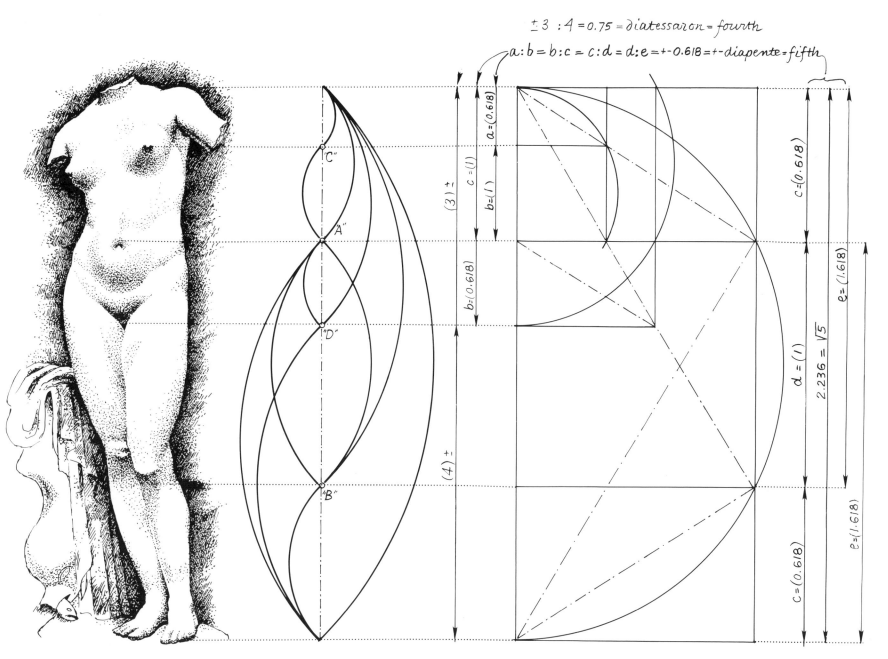

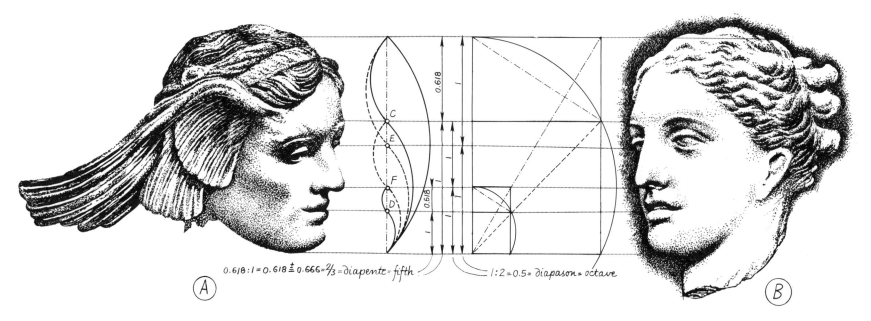

Fig. 153. **A.** Hypnos, goddess of sleep. **B.** Hygeia, goddess of health.

The heads of Hypnos, goddess of sleep (fig. 153-A) and of Hygeia, goddess of health and patroness of the Pythagoreans, (fig. 153-B) both from the fourth century B.C., share in miniature the same proportional limitations which articulate the bodies of the Cyrenian Aphrodite and the Doryphoros. Scales and wave diagrams beside these heads reveal how the vertical distances between the main features from top of head to chin share the same simple numeric relationships found earlier in all living shapes. The Greeks thought that humans, with all their limitations, had the capacity to reflect limitless harmony and beauty, conceived as divine. Thus man was said to be the measure of all things.

In fact, we are told by Vitruvius, the Roman architect mentioned in the previous chapter, that the ancient Greeks even laid out their temples according to human proportions. On this basis Vitruvius recommends that the length of a temple should be twice its width, and the proportions of the open entrance hall (pronaos) and the enclosed inner room (cella) should be in 3-4-5 relation (3 being the depth of the pronaos, 4 the width, and 5 the depth of the cella). Figure 154 shows these proportional relationships as they correspond to musical harmonies.

Vitruvius also furnished many other recommended temple proportions, all based upon Greek models, for instance regarding the distances between columns and their proper height, both expressed in terms of column diameter. Such an element, chosen in order to express the proportions of an entire structure (as the proportions of the human body are expressed in feet) is called a *module,* a concept that was to play an important role throughout the history of architecture.[72]

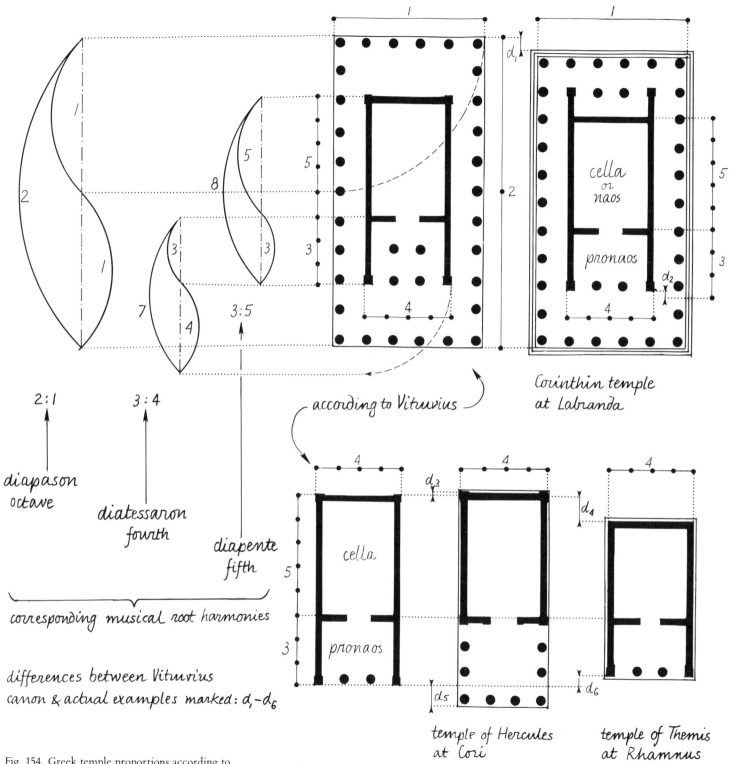

2:1

3:4

3:5

diapason
octave

diatessaron
fourth

diapente
fifth

corresponding musical root harmonies

differences between Vitruvius
canon & actual examples marked: d_1–d_6

according to Vitruvius

Corinthin temple
at Labranda

temple of Hercules
at Cori

temple of Themis
at Rhamnus

Fig. 154. Greek temple proportions according to
Vitruvius, compared with actual examples.

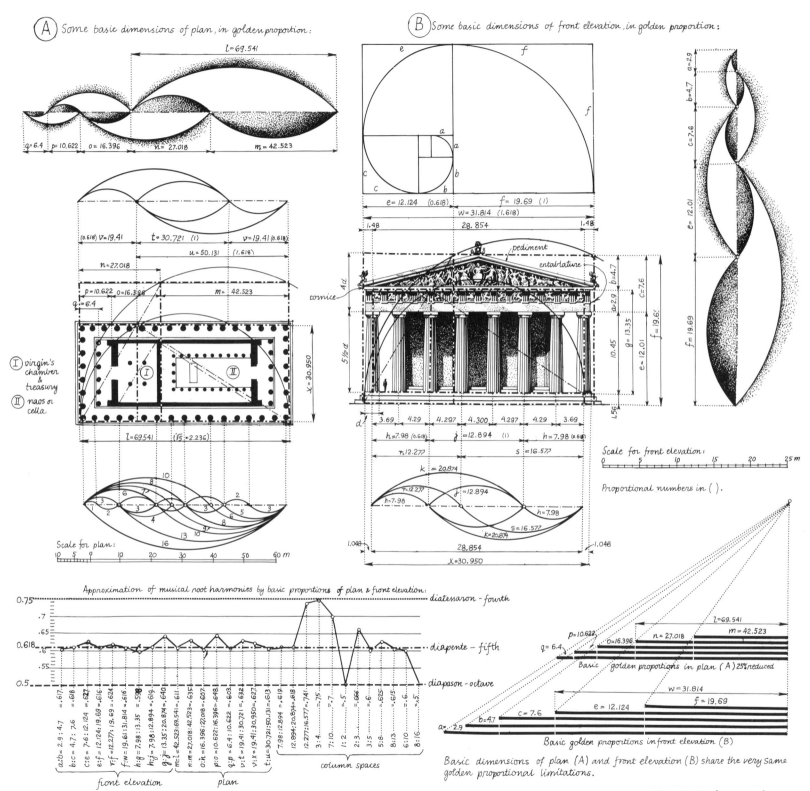

A Some basic dimensions of plan, in golden proportion:

L= 69.541
q= 6.4   p= 10.622   o= 16.396   n = 27.018   m = 42.523

(0.618) V=19.41   t= 30.721 (1)   v= 19.41 (0.618)
u= 50.131 (1.618)
n= 27.018
p= 10.622 o= 16.396   m = 42.523
q= 6.4

I virgin's chamber & treasury
II naos or cella

l= 69.541   (√5 = 2.236)

x= 30.950

Scale for plan:
10 5 0   10   20   30   40   50   60 m

B Some basic dimensions of front elevation, in golden proportion:

e   f

f

a
a

c   b
c   b

e= 12.124 (0.618)   f= 19.69 (1)
W= 31.814 (1.618)
1.48   28.854   1.48

pediment
entablature
cornice

a= 2.9 b= 4.7 c= 7.6
g= 13.35
e= 12.01
f= 19.6(

10.45

1.56

d   3.69 4.29 4.297 4.300 4.297 4.29 3.69
h= 7.98 (0.618)   j =12.894 (1)   h= 7.98 (0.6(8)
r=12.277   s = 16.577

k = 20.874
r=12.277   j =12.894
h= 7.98   h= 7.98
s = 16.577
k= 20.874
1.048   28.854   1.048
X=30.950

Scale for front elevation:
0   5   10   15   20   25 m

Proportional numbers in ( ).

a= 2.9
b= 4.7
c= 7.6
e= 12.01
f= 19.69

Approximation of musical root harmonies by basic proportions of plan & front elevation:

0.75 ———————————————————— diatessaron - fourth
.7
.65
0.618 ———————————————————— diapente - fifth
.6
.55
0.5 ———————————————————— diapason - octave

a:b= 2.9 : 4.7 = .617
b:c= 4.7 : 7.6 = .618
c:e= 7.6 : 12.124 = .627
e:f= 12.124:19.69 = .616
r:f= 12.277 : 19.69 = .624
f:w= 19.61:31.814 = .616
h:g= 7.98 : 13.35 = .598
h:j= 7.98 : 12.894 = .61.9
g:j= 13.35 : 20.874 = .640
m:l= 42.523:69.541 = .611
n:m= 27.018 : 42.523 = .635
o:n= 16.396:27.018 = .607
p:o= 10.622 :16.396 = .648
q:p= 6.4 : 10.622 = .603
v:t= 19.41:30.721 = .632
v:x= 19.41: 30.950 = .627
t:u= 30.721:50.131 = .613
7.98 : 12.894 = .619
12.894 : 20.874 = .618
12.277 :16.577 = .741
3: 4 = .75
7:10 = .7
1 : 2 = .5
2 : 3 = .666
3:5 = .6
5:8 = .625
8:13 = .615
6:10 = .6
8:16 = .5

column spaces

front elevation   plan

Basic golden proportions in plan (A) 25% reduced

l= 69.541   m = 42.523
p= 10.622 n = 27.018
q= 6.4   o= 16.396

W= 31.814
e = 12.124   f= 19.69
a = 2.9 b= 4.7 c= 7.6

Basic golden proportions in front elevation (B)

Basic dimensions of plan (A) and front elevation (B) share the very same golden proportional limitations.

Fig. 155. Parthenon, Athens.

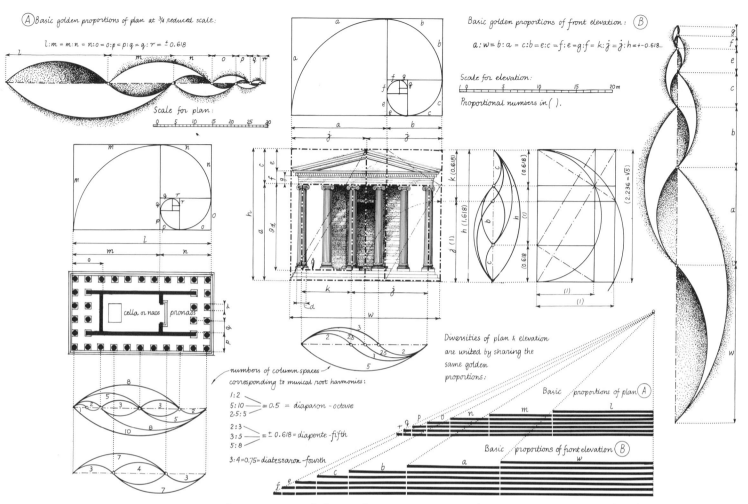

(A) Basic golden proportions of plan at ¾ reduced scale:

$$l:m = m:n = n:o = o:p = p:q = q:r = \pm 0.618$$

Scale for plan:

(B) Basic golden proportions of front elevation:

$$a:w = b:a = c:b = e:c = f:e = g:f = k:j = j:h = \pm 0.618...$$

Scale for elevation:

Proportional numbers in ( ).

cella or naos   pronaos

numbers of column spaces corresponding to musical root harmonies:

1:2
5:10 ⟶ = 0.5 = diapason - octave
2.5:5

2:3
3:5 ⟶ = ± 0.618 = diapente - fifth
5:8

3:4 = 0.75 = diatessaron - fourth

Diversities of plan & elevation are united by sharing the same golden proportions:

Basic proportions of plan **(A)**

Basic proportions of front elevation **(B)**

Fig. 156. Athena temple, Priene.

The recommended temple proportions can be seen in two examples, belonging to two different styles: the Parthenon of Athens (fifth century B.C.), (fig. 155) representing the Doric order, and the Athena temple of Priene (fourth century B.C.), embodying the Ionic style (fig. 156). The difference between the two styles is exemplified by the columns: the Doric being more sturdy, the Ionic more slender. The height of the Parthenon's columns is five and a half times the width of the column base. The capitals consist of simple square slabs (abaci) resting upon shapes, the contours of which have been likened to two outstretched hands (echini). The height of the Athena temple's columns is nine times the width of the column base, and the Ionic capital is graced by two shell-shaped volutes.

The Parthenon's main facade fits into a single reclining golden rectangle, while the facade of the Athena temple rises within two upright ones. Relations of superstructure to supporting columns plus steps echoes the same proportions in two variations. In the Parthenon, the top of the column capitals approximates the golden section point of the total height, while in the Athena temple this point corresponds to the meeting line between two reciprocal golden rectangles. In the Parthenon, the centerlines of the two corner columns and the lines of the floor and entablature-top form a $\sqrt{5}$ rectangle, consisting of two reciprocal golden rectangles, while in the Athena temple, three neighboring column-centerlines encompass a single golden rectangle. (See dotted diagonals in elevations.)

The series of columns in themselves contain proportional rhythms, the columns and the spaces between them representing a "recurrence of strong and weak elements," a standard definition of rhythm. The front columns of the Parthenon with their seven spaces embody the 3:4 ratio of the Pythagorean triangle and the corresponding musical harmony of fourth-diatessaron, as well as approximations of golden proportions or fifth-diapente harmonies. In the Athena temple, 2:3 and 3:5 relations of columns approximate the root harmony of diapente.

The rhythmic diagrams below the columns express some of these latent equivalents of musical root harmonies also along the long sides of the buildings. The Athena temple is about twice as long as it is wide, bearing out Vitruvius and embodying the 1:2 harmony of octave-diapason. The Parthenon's plan corresponds to two reciprocal golden rectangles, thus echoing diapente harmony. The interior structures of both temples also confirm Vitruvius, at least partially. The pronaos of the Athena temple is in 3:4 proportion, while the naos or cella in both temples, and the treasury or virgin's chamber in the Parthenon are in golden proportion.

It is fascinating to discover the unity of proportional limitations—corresponding to patterns of organic growth as well as to root harmonies of music—in the diversities of these temples, both in plan and in elevation. Diagrams **A** and **B** in figs. 155 and 156 illustrate this by showing that characteristic dimensions of both plans and elevations can be represented by the same diagrams that were used earlier for patterns of organic growth. Organ-pipe-like bar diagrams, representing the same dimensions, further illustrate this proportional unity, by the fact that all of these main articulations converge in a single point, as seen earlier in the anatomy of the horse.

The goddess Athena, to whom the Parthenon was also dedicated, united in her person both "masculine" and "feminine" virtues. She was the patroness of wisdom and of the arts and crafts and she was an example of valor in battle. The grace of her Priene temple and the strength of the Parthenon express the unity of these diverse human qualities in her, a divine model of human wholeness.

The diversity between the Romans and the Greeks becomes apparent from the kind of buildings in which they excelled. Whereas the Greeks built temples and theaters of unsurpassed beauty, the Romans built roads, aqueducts, palaces, public bath-houses, triumphal arches and circuses, coupling the beauty of the Greeks with a mastery of engineering techniques. They added arches, vaults and domes to the columnal orders inherited from the Greeks, and they increased everything to a colossal scale, making the Greek temples look modest.

In spite of these differences, Greek and Roman architecture are united by identically shared proportional limitations. Two examples may illustrate this: the Triumphal Arch of Constantine, (fig. 157) and the Colosseum (fig. 158) both in Rome.

The Triumphal Arch of Constantine is a veritable treasury of golden relationships. The overall shape of this structure approximates two golden rectangles (**d:e**). Half the width projected upon the height establishes the springline of the central vault (**A**) as well as the bottom line of the architrave (**B**), (see diagrams and golden section constructions at right).

The top of the main cornice (**C**) as well as the springline of the small archways (**D**) coincide with the sides of two golden rectangles flanking a square in a semicircle, the classical construction of the golden section. A diagram (lower left) shows that the heights of the major archway and of the minor ones (**o** and **p**) as well their difference (**n**) and two articulations of the architrave (**m** and **l**) form a series of golden relationships which again coincide with the pattern

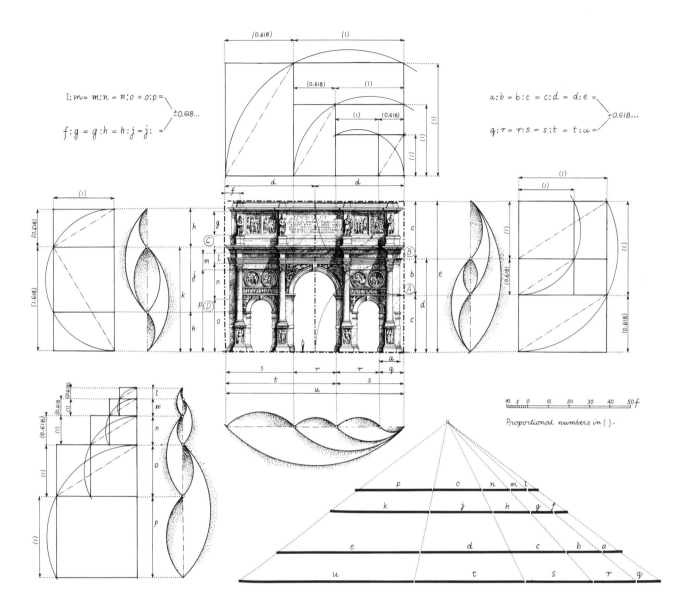

Fig. 157. Triumphal Arch of Constantine, Rome.

of organic growth. Diagrams above and below the elevation indicate how the golden section relates to each other all major articulations also in horizontal direction.

Proportional study of the Colosseum shows that the plan fits into two golden rectangles (**b:a**) and that the width of the giant ellipse which forms the outside wall relates to the width of the central arena in the $\sqrt{5}$ proportion of two reciprocal golden rectangles (see diagram and construction above plan). Similar relationships exist along the shorter axis of the plan (see diagrams and constructions at the right of the plan).

The same relationships reappear in elevation between the heights of the three lower stories (**n:o:p**) and also between these latter combined and the crowning fourth story (**r:s:t**) (see diagrams and construction at right of elevation). All these proportions approximate the musical root harmony of the fifth-diapente, while the central location of the cornice over the second floor corresponds to octave-diapason.

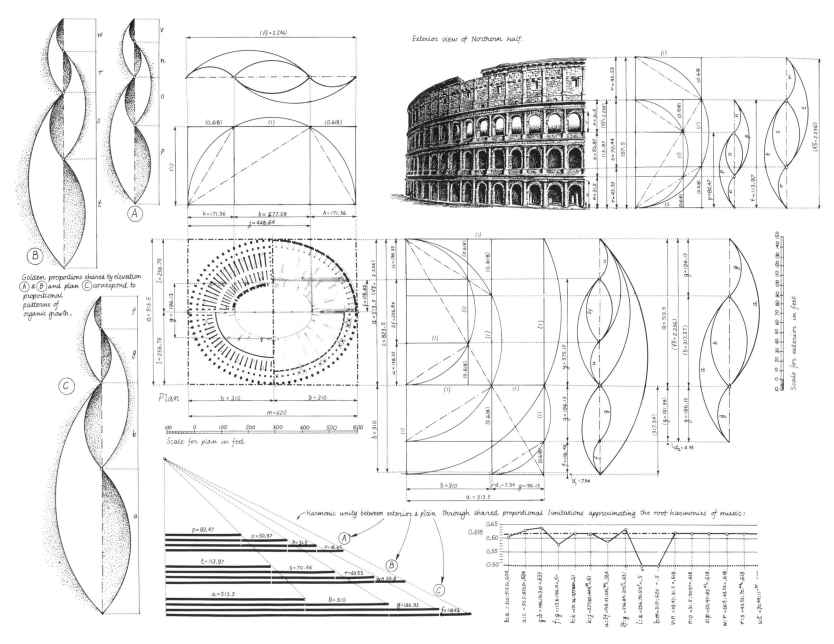

Fig. 158. Colosseum, Rome.

Rhythmic wave diagrams **A, B,** and **C** illustrating basic proportions of plan and elevation reveal that these again correspond to organic growth patterns. Harmonic unity achieved between the diversities of plan and elevation, through shared proportional limitations is again demonstrated by bar diagrams, all articulations of which converge to the very same point.

Roman poets apparently shared the Roman architect's devotion to the proportions of the golden section. According to Professor G. E. Duckworth of Princeton University, these proportional relations can be found in the poetry of Catullus, Lucretius, Horace, and Virgil.[73] For instance, in Virgil's celebrated epic, *The Aeneid,* the number of lines of the various parts of the narrative are never arbitrary, but always approximate golden relationships, often corresponding to Fibonacci numbers.

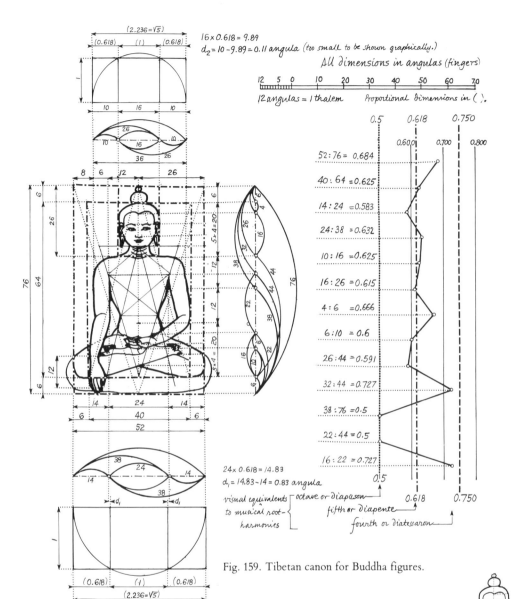

Fig. 159. Tibetan canon for Buddha figures.

Measuring the immeasurable

The differences between Greek and Roman art dwindle when we compare these classical forms to the arts of the East. The five examples we will look at were chosen for their diversity as well as for their importance within their cultures. As we will see, the proportions we have found in the West also appear in the arts of the East.

The Tibetan canon for the construction of Buddha figures, published in Benjamin Rowland Jr.'s book, *The Evolution of the Buddha Image* in 1976, shows how the figure fits into three golden rectangles within each other.[74] (fig. 159) The largest rectangle encloses the whole figure from top knot to base, encompassing the knees; a smaller one reaches from top of head to legs, touching right hand and elbow; and the smallest frames the head. Wave diagrams and constructions of the golden section above and below the figure reveal reciprocities that unite the upper and lower portions of the body, suggesting the compassion of the supreme Buddha toward even the lowest living things. Corresponding musical root harmonies are shown by the vertical wave diagram and the graph.

The two triangles which appear in the original Tibetan canon, reaching from chin to legs, are shown to correspond with diagonals of half the enclosing golden rectangles, delineating a central pentagon and pentagram pointing to chin, waist, and armpits.

The Korean bronze figure of Maitreya, the future Buddha (fig. 160), can be enclosed in a golden rectangle resting upon a square of equal width, the latter containing the seat, the

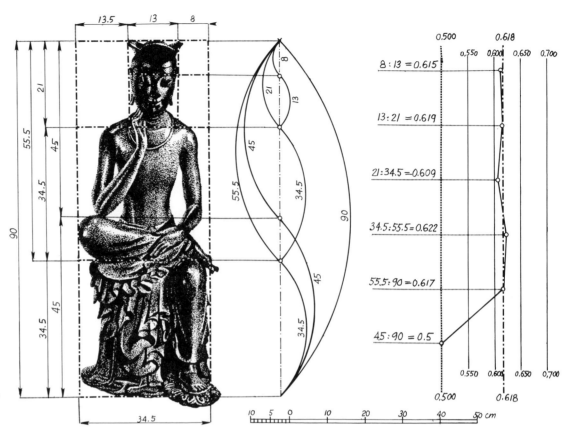

Fig. 160. Figure of Maitreya.

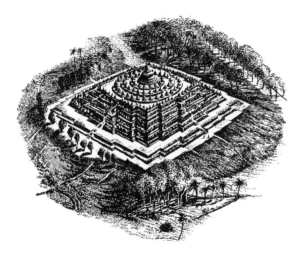

former framing the bulk of the figure itself. Wave diagrams and graph show that the critical point, where the right elbow rests upon the right knee, is exactly at the center of the whole height, corresponding to diapason musical harmony. Other important relationships, such as shoulder height between top of head and seat, and forehead height between top of head and shoulder, approximate golden proportions and thus the harmony of diapente.

The similarity is striking between these proportional relationships and those prevailing in Greek art. It is well known that historic contacts existed between classical Western art and Buddhist art, through trade and the conquests of Alexander the Great in the fourth century B.C. This explains the physical means by which information may have been transmitted, but the universal appeal of these particular proportions surely springs from other sources. This universality is further demonstrated by a look at Buddhist architecture.

Borobudur, the world's largest Buddhist stupa (fig. 161), was built at the end of the eighth century on the island of Java, Indonesia. It consists of eight stupa-covered terraces arising from a raised base that covers about one and a half city blocks (115 meters). The five lower terraces are square, the three upper ones are circular: 3-5-8 are, of course, Fibonacci numbers.

Buddha figures are enshrined in niches of the bas-relief-covered balustrades that surround the square terraces. Larger bell-shaped stupas on the circular terraces house larger Buddha figures, all representing various stages of enlightenment. Even the numbers of these stupas—32-24-16— approximate musical root harmonies in their relationship to each other, as the graph indicates. At the center of the top terrace is the greatest stupa containing the largest Buddha, representing the highest attainment.

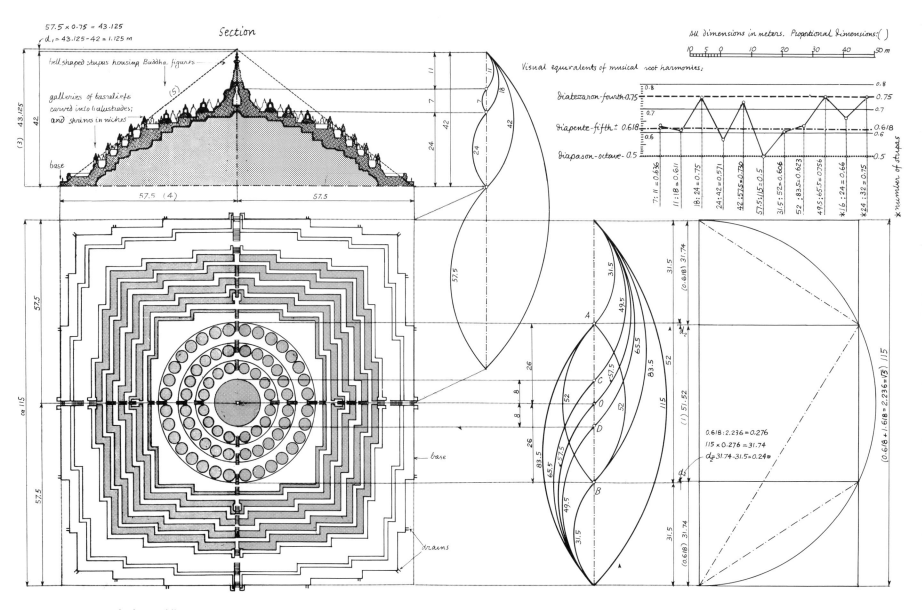

Fig. 161. Borobudur Buddhist stupa, Java.

The diameter of the largest circular terrace and the width of the square base are again in golden relationship with each other, indicated by points **A** and **B** in the wave diagram at the right, and by the golden section's classical construction, containing two reciprocal golden rectangles (dash-dotted lines). The section of the Borobudur stupa shows how the height of the supreme stupa and the base line form two 3-4-5 triangles, similar to the Mexican Pyramid of the Sun and the Great Pyramid of Egypt, though in the latter the two triangles stand on their short sides. Wave diagrams and graph indicate further harmonious proportions, approximating musical root harmonies.

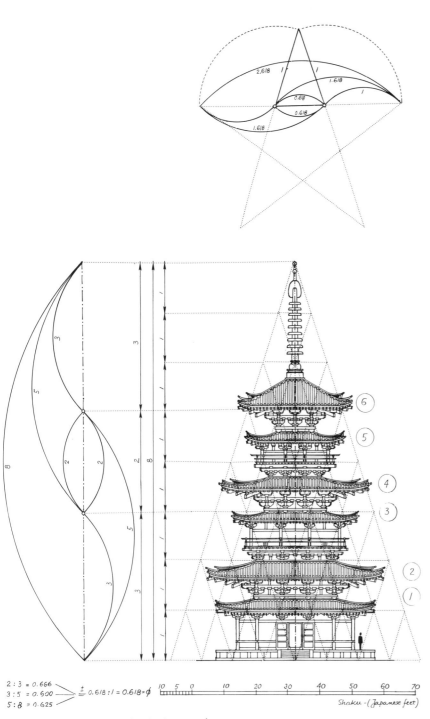

2:3 = 0.666
3:5 = 0.600
5:8 = 0.625

± 0.618:1 = 0.618 = φ

Shaku - (Japanese feet)

Fig. 162. Pagoda of Yakushiji temple.

In Japan, one finds entirely different forms of Buddhist architecture, exemplified by the Pagoda of the Yakushiji temple, known for its soaring grace as well as for its ingeniously contrived structural strength. (fig. 162). It is built of wood and consists of six roofs, all of different sizes, rising one above the other, culminating in a tall steeple. Above the ground floor there are two balconied floor levels, and each of the floors has two roofs. The entire structure articulates into eight equal heights. Of these, two contain the ground floor, two the steeple, and the remaining four contain the two upper floors in between. The heights of the three lowest and the three highest levels relate to the heights of the two central ones in proportions approximating golden ones, as do the five lowest to the three highest and vice versa.

The prevalence of these reciprocal proportions in the entire structure is also shown by the fact that its overall outline fits neatly into one of the pentagonal star's triangles, and it may be recalled that all relationships in this star pattern are golden ones. A network of minor pentagram triangles interrelates many salient points of this structure, such as the ridgelines of the second and third roofs; the length of the first, second, and fourth; the baseline of the steeple and the eavelines of the highest and the lowest roofs.

The wave diagrams reveal how waves of proportional relationships pulsate through the entire body of the building, as if it were a living organism. Notations of these relations are shown in the wave diagrams: "**m**" stands for minor, "**M**" for major members of a golden relation, "**M̲**" designating the combination of the two, and "**M̲**" referring to a still further combination, such as the $\sqrt{5}$ rectangle, containing two reciprocal golden relationships. This notation shows at a glance how minor elements in one relationship become major ones within the next, as in processes of organic growth. The letters "**t**" and "**T**" (for tessaron) indicate proportions close to the 0.75 rate of diatessaron and also to the Pythagorean triangle's 3:4 relations, while "**o**" and "**O**" stand for the 0.5 or 1:2 relation of octave or diapason.

The heavier wavelines of the vertical wave diagram show two golden section points of the total height—one at the ridge of the third roof **A**, the other at the eave of the sixth roof **B**. Lighter wave lines indicate some of the many similar relationships among diverse other heights.

It is interesting to observe how vertical proportions are shared by horizontal ones. The floor and roof structures share these relationships in pairs: **1** with **6** (the highest and lowest), **2** with **3**, and **4** with **5**. Within each of these parts of the structure one again finds correspondences to all three musical root harmonies, both in the overall shapes and in more minute details, including height to width, length of overhang to supporting structure, steeple height to height and width of top roof, and so on.

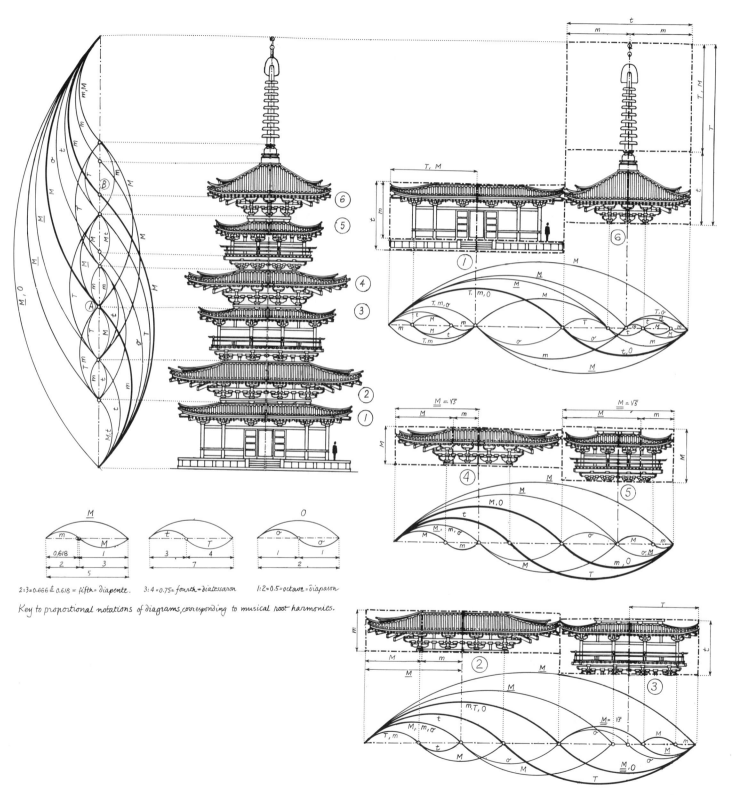

2:3=0.666 ≒ 0.618 = *fifth = diapente*. 3:4=0.75= *fourth = diatessaron* 1:2=0.5=*octave = diapason*

Key to proportional notations of diagrams, corresponding to musical root harmonies.

In entirely different ways, the same proportional harmonies can be discovered in the Ryoan-ji Zen temple's garden, near Kyoto, which dates from the beginning of the fifteenth century. (fig. 163) This garden is designed to be seen from the veranda of the monastery and from the paved walkways around it. The rectangle of coarse, raked white sand is never walked on. There are five groups of rocks on the sand apparently at random, like rocks along the shoreline of the sea. The rocks are placed on the sand apparently at random, yet they seem to belong together as naturally as the land and the sea. One secret of this relatedness is revealed by the subtle proportions that unite the overall shape of the garden with the distances between rocks and enclosure.

The field of sand has proportions corresponding to two reciprocal golden rectangles, as shown by dash-dotted diagonals which are identical with those of the construction of two reciprocal golden rectangles, below the plan. Lines **A** and **B** connect the rocks. Line **B** is identical with the diagonal of one of the reciprocal golden rectangles constituting the sand field. Line **A** connects the three-quarters point of the field's eastern side with point **C** along the opposite side, this latter point corresponding to the meeting of two reciprocal golden rectangles, as the construction demonstrates. Wave diagrams and multiple constructions of the golden section above lines **A** and **B** show how the distances between the rocks within the field share proportional relationships corresponding to the root harmonies of music. Thus rocks and field become one.

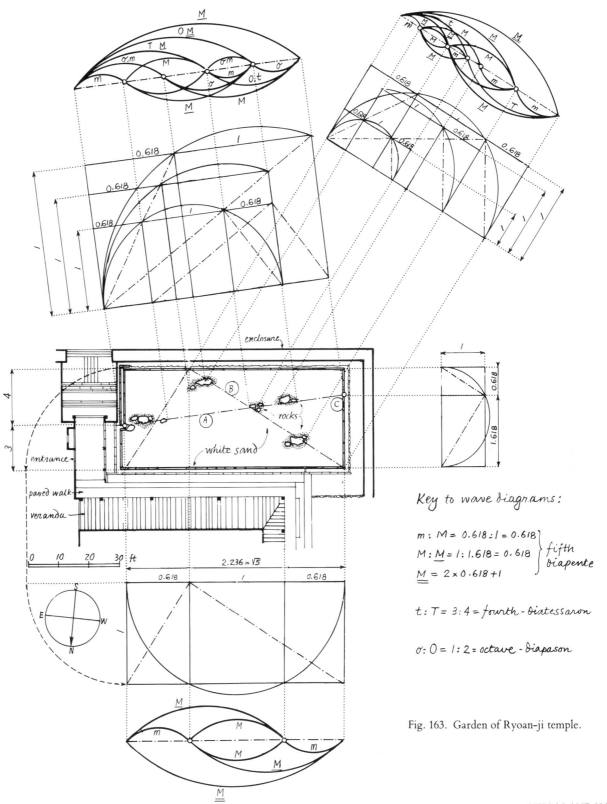

Key to wave diagrams:

$m : M = 0.618 : 1 = 0.618$ ⎫
$M : \underline{M} = 1 : 1.618 = 0.618$ ⎬ fifth diapente
$\underline{M} = 2 \times 0.618 + 1$ ⎭

$t : T = 3 : 4 =$ fourth - diatessaron

$\sigma : O = 1 : 2 =$ octave - diapason

Fig. 163. Garden of Ryoan-ji temple.

We now come to Japan's oldest, and at the same time newest, architectural gem, the Ise Shrine's East Treasure House, dedicated to the goddess of food, Toyo-uke-bime-no-kami. (fig. 164) It is the oldest because it dates back to the fourth century, built according to prototypes of raised prehistoric dwellings. It is also the newest because it is torn down and rebuilt every twenty years, in order to preserve its original integrity, always faithfully constructed following the same design, one unsurpassed in its clarity. All parts retain their natural shapes and finishes: round posts are of white cedar, grown near the site, the bark of the same trees is used for the roof; and the beams, rafters, and wall planks are unpainted, revealing the natural beauty of the wood grain.

The plan of the building is a single golden rectangle. The gable elevation, from ground to ridgeline and from eave to eave, corresponds to two golden rectangles. The side elevation,

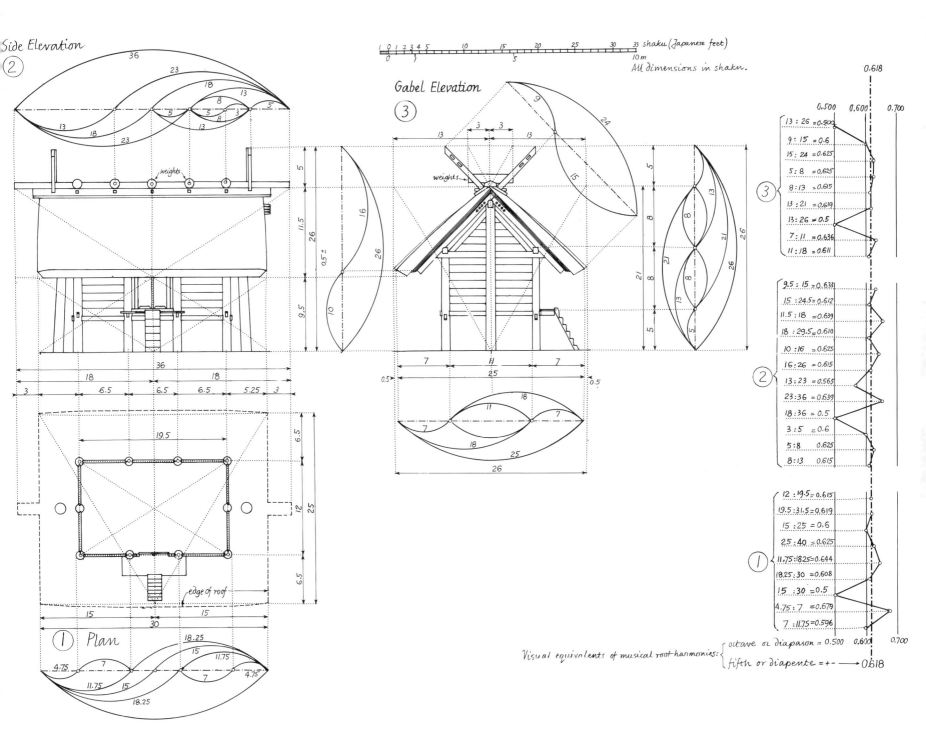

Fig. 164. Ise Shinto Shrine's East Treasury House.

between ground and eaveline and between the edges of the roof, fits again two golden rectangles, topped by two larger ones that cover the roof itself. Nothing is arbitrary; these same proportional limitations exist between roof overhangs and the building proper, as well as between the windmill-like extensions of the rafters and the roof itself, shown at a glance by the wave diagrams around the gable elevation.

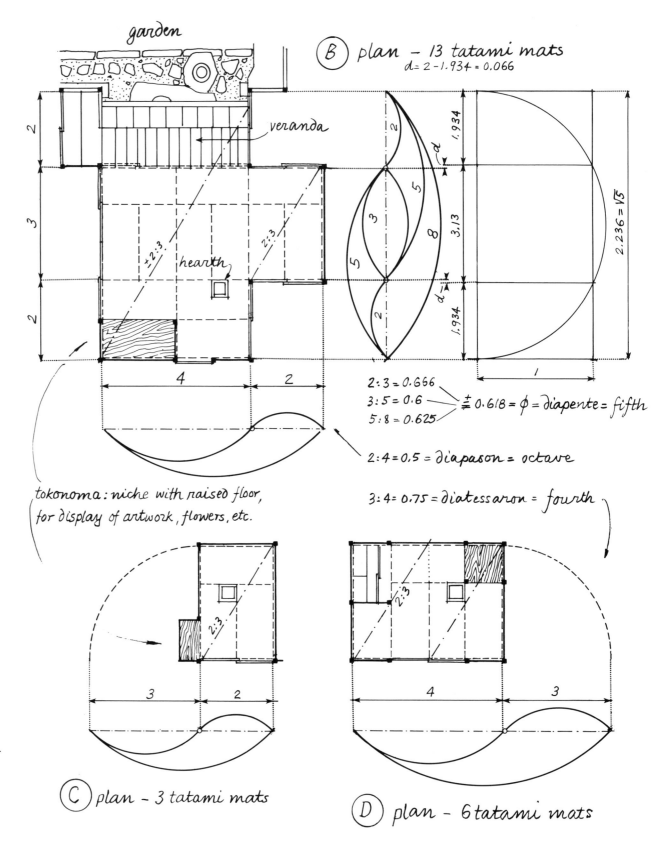

garden

Ⓑ plan – 13 tatami mats
$d = 2 - 1.934 = 0.066$

veranda

± 2:3

hearth

2:3

2
3
2

4
2

2
5
3
5
8
2

1.934
d
3.13
1.934

2.236 = √5

1

$2:3 = 0.666$
$3:5 = 0.6$
$5:8 = 0.625$
$\underset{=}{\pm} 0.618 = \phi = $ diapente = fifth

$2:4 = 0.5 = $ diapason = octave

$3:4 = 0.75 = $ diatessaron = fourth

tokonoma: niche with raised floor,
for display of artwork, flowers, etc.

Ⓒ plan – 3 tatami mats

2:3

3
2

Ⓓ plan – 6 tatami mats

2:3

4
3

Fig. 165. Proportions of Japanese tea rooms.
A, B: Bosen (final attainment) tea rooms.
C, D: Typical tea room plans.

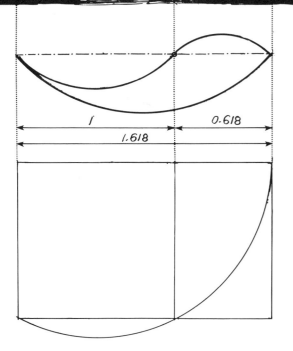

A view over veranda into garden

The greatness of little things

The Japanese tea ceremony is the celebration of the simple joys of life. This ritual evolved from traditions of Zen Buddhism, which teaches the value of limits by the discovery of "greatness of little things."[75] All the teahouses and tea gardens in Japan were laid out by revered tea masters. They were often painters, poets, architects, and gardeners as well, like Koburi Enshu, who built the Bosen (final attainment) tearoom in the seventeenth century. (fig. 165) It is one of the earliest examples of the union of inside and outside, a feature which became important in the modern architecture of the west. Here it is accomplished by means of suspended shoji screens, made of rice paper pasted on a light wooden lattice. As the drawing shows, the proportions of this shoji and its opening towards the veranda correspond to reciprocal golden relations—a tangible reciprocity that unites the intangible diversities separating man from nature.

As in most Japanese houses, the floor is spread with tatami mats, made of straw and covered with woven rush, usually three feet wide and six feet long, thus incorporating the 1:2 harmony. The floor plan **B** reflects golden proportional harmonies. The main room and veranda jointly correspond to a golden rectangle (see dash-dotted diagonal marked **2:3**), and the extension of the room, three tatami mats, share the same proportions. Figures **C** and **D** show how the equivalents of the musical root harmonies 2:3 and 3:4 are embodied in the plans of two typical smaller tearooms. It is noteworthy that the pentatonic music of the East (and much folk music in the West) is limited to five tones, thus remaining always close to the root harmony of fifth-diapente.

The use of tatami mats ensures that in Japanese architecture room sizes and also the grouping of rooms into entire buildings will always be in harmonious proportions. The tatami mats thus act as modules for all floor layouts, just as the shoji screens act as modules for the vertical

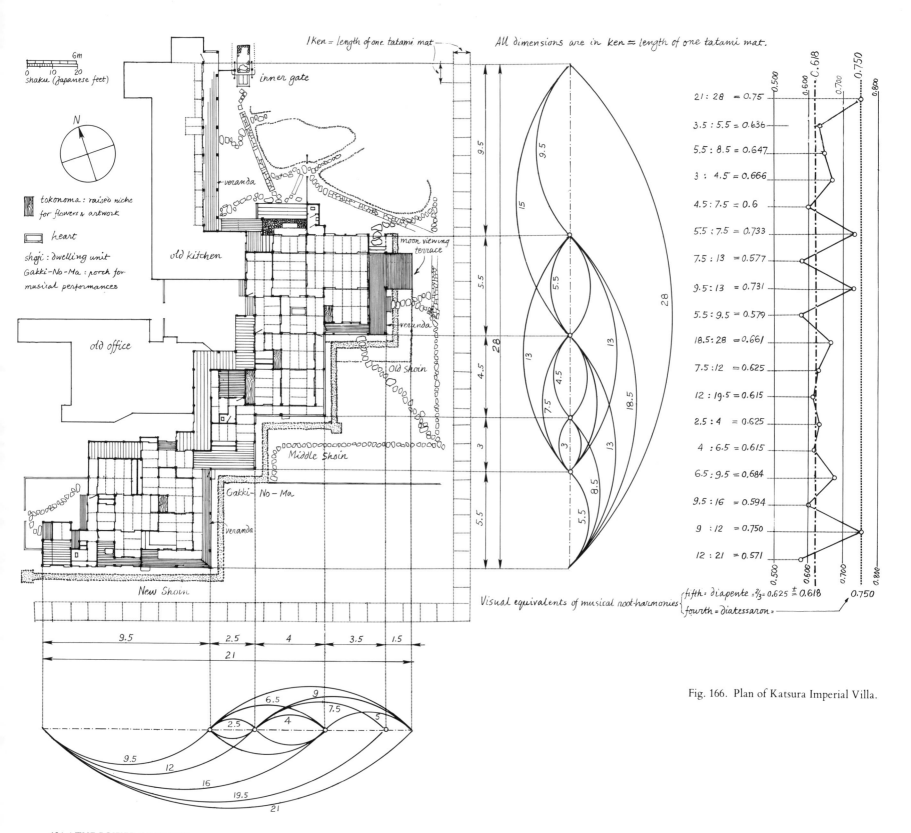

All dimensions are in ken = length of one tatami mat.

1 Ken = length of one tatami mat

6m
shaku (Japanese feet)

N

tokonoma : raised niche
for flowers & artwork

hearth

shoji : dwelling unit
Gakki-No-Ma : porch for
musical performances

inner gate

veranda

old kitchen

old office

moon viewing
terrace

veranda

Old Shoin

Middle Shoin

Gakki- No- Ma

veranda

New Shoin

9.5   2.5   4   3.5   1.5

21

2.5   6.5   9
       4   7.5   5

9.5
       12
              16
                     19.5
                            21

Visual equivalents of musical root-harmonies { fifth: diapente = ⅔ = 0.625 ± 0.618 / fourth = diatessaron = —

21 : 28 = 0.75
3.5 : 5.5 = 0.636
5.5 : 8.5 = 0.647
3 : 4.5 = 0.666
4.5 : 7.5 = 0.6
5.5 : 7.5 = 0.733
7.5 : 13 = 0.577
9.5 : 13 = 0.731
5.5 : 9.5 = 0.579
18.5 : 28 = 0.661
7.5 : 12 = 0.625
12 : 19.5 = 0.615
2.5 : 4 = 0.625
4 : 6.5 = 0.615
6.5 : 9.5 = 0.684
9.5 : 16 = 0.594
9 : 12 = 0.750
12 : 21 = 0.571

Fig. 166. Plan of Katsura Imperial Villa.

articulation of the walls. The plan of the Katsura Imperial Villa in Kyoto (fig. 166), which is a free combination of diversely shaped and sized rooms, shows how the use of the modular tatami mats creates a rhythmic and harmonious unity and wholeness, without becoming monotonous or forced. Thus the disciplined order of shared proportional limitations achieves freedom and grace. The proportions of each room correspond to the root harmonies of music, as the wave diagrams and the graph demonstrate. Music played an important part in the conception of this plan: music being played on the veranda of the Middle Shoin could be enjoyed from the veranda of the Old Shoin while watching the rising of the moon.

The exterior of the Katsura Villa (fig. 167) harmonizes with the plan through its own golden proportions, as can be seen from drawing **A,** which reveals that one of the main openings relates to its two flanking shoji panels as the square relates to its two flanking golden rectangles in the classical construction of the golden section. Similarly, construction **B** shows how the proportions of the openings and shoji screens are shared by the spacing of the structural columns, integrating thereby the rhythmic harmony and wholeness of the entire building.

The shared proportional limitations of architecture reflect a basic pattern-forming principle permeating all Japanese life and art. For instance, haiku poems are limited to seventeen syllables in three lines of five, seven, and five syllables. These narrow limits allow powerful expression through suggestion by detail, a form of power of limits.

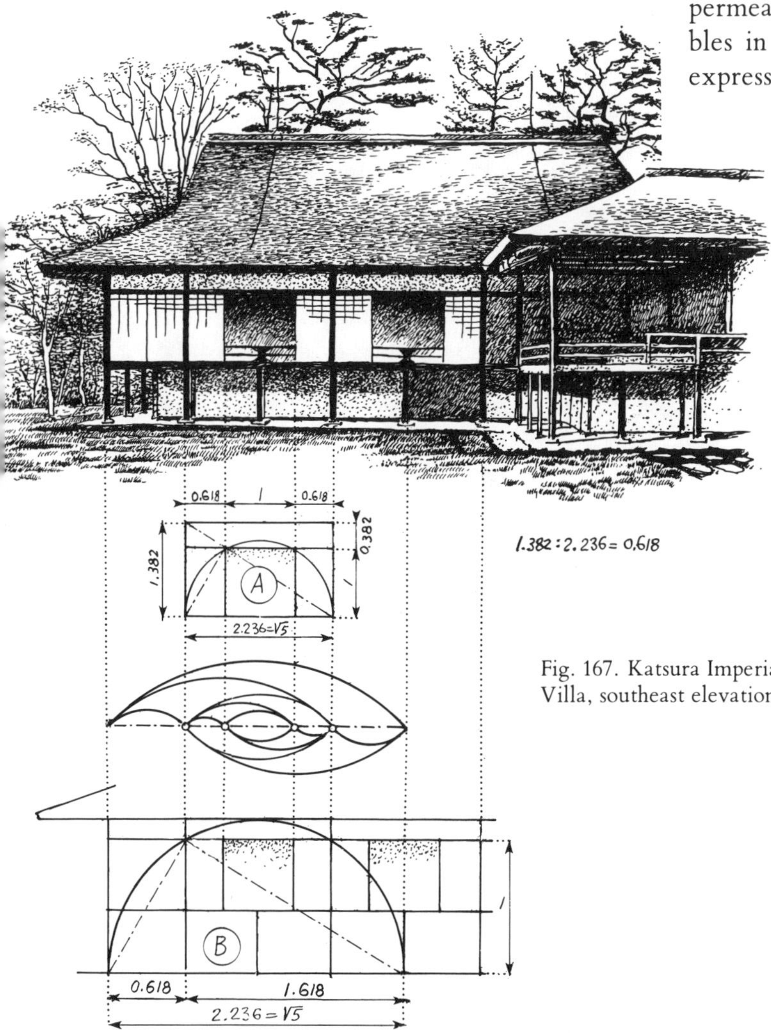

$1.382 : 2.236 = 0.618$

Fig. 167. Katsura Imperial Villa, southeast elevation.

The piercing chill I feel:
my dead wife's comb, in our bedroom,
under my heel.

    —Buson, 18th century

A mountain village:
Under the piled-up snow
the sound of water.

    —Shiki, 19th century

Fallen petals rise
back to the branch—I watch:
oh. . .butterflies!

    —Moritake, 15th century

I have just come from a place
at the lake bottom! That is the look
on the little duck's face.

    —Joso, 17th-18th century[76]

Fig. 168. Chinese monk at the
moment of enlightenment.

CHAPTER 8: Wisdom and Knowledge

Eastern and Western arts of living

Among the essentials needed to turn mere survival into the art of living, perhaps none are more important than wisdom and knowledge. In a certain sense, these two human aptitudes are almost indistinguishable from each other; in another sense they are polar opposites. Wisdom is a putting together, knowledge a taking apart. Wisdom synthesizes and integrates, knowledge analyzes and differentiates. Wisdom sees only with the eyes of the mind; it envisions relationship, wholeness, unity. Knowledge accepts only that which can be verified by the senses; it grasps only the specific and the diverse.

The origins of these words hint at their opposition. Wisdom comes from the Indo-European root verb *weid,* "to see," the same root from which words like "vision" and "Veda" come, the latter being the name of the ancient, sacred teachings of India, meaning literally "I have seen." Knowledge, on the other hand, originates from the root *gno,* "to know," which gave birth also to the words "can" and "cunning."

Both wisdom and knowledge are based on experience, but wisdom more so than knowledge, which frequently retains experience only through the filter of conceptual thought, at times discarding the seeds of life. In contrast, wisdom often stammers, or speaks in images, symbols, paradoxes, or even riddles.

One could say that the greatness of the East has been its dedication to wisdom, while the West has been concentrating, particularly in the last century or two, upon knowledge. This emphasis in the West has brought about the phenomenal flourishing of science and technology, but unfortunately without a similar development of wisdom, though of course the roots of Western culture are not lacking in ancient wisdom of their own, as we have seen. But knowledge and wisdom are equally essential dinergic diversities that should complement each other. A dinergic countermove to supplement our present one-sidedness is the increasing Western interest in the teachings of the Orient.

The Nobel-prize-winning physicist Erwin Schrödinger said in 1956 in his lectures at Trinity College, Cambridge: ". . . our present way of thinking does need to be amended perhaps by a bit of blood-transfusion from Eastern thought."[77] This blood-transfusion is actually in progress, in the form of Western studies of Eastern medicine, yoga, Buddhism, Tai Chi, and other Oriental disciplines, as well as such books as R.G.H. Siu's *The Tao of Science: An Essay on Western Knowledge and Eastern Wisdom,* to which this chapter is indebted.

Following are a few highlights of Eastern wisdom compared with some salient features of Western knowledge, in an attempt to show that they are indeed dinergic diversities sharing the same basic pattern-forming processes. If we could unite them, we and our world would become more whole.

A reflection of the compassionate teaching of the Buddha was revealed by the reciprocally related higher and lower elements in the Tibetan canon of his figure, by similar reciprocities between the highest and lowest terraces of the Borobudur stupa, as well as by the highest and lowest rooflines in the pagoda of the Yakushiji temple. These teachings were preceded in India by the wisdom of the Vedas, expressed as *tat twam asi*, "this thou art," teaching the relatedness of all things and the nonviolent attitude that results from that realization.

The unity of diversities in the individual, between one's consciousness and that which is beyond it, is the key experience of Zen Buddhism, reflected, for instance, by the complete reciprocities between rocks and sand field in the Ryoanji Zen garden. The flash of this unifying experience, called *satori* ("enlightenment" or "insight" in Japanese), was captured six hundred years ago in the sculpture of a Chinese monk (Japanese Zen having developed from Chinese Ch'an Buddhism). (fig. 168)

The Buddha taught about the need to avoid excess by treading the Middle Path between self-indulgence and self-mortification. Thus he expressed in terms of human behavior the harmony of the golden mean. The bliss of wholeness found in the enlightenment of the Middle Path is often described as new birth, frequently symbolized by the unfolding of the mind's thousand-petalled lotus.

The great Chinese teacher Confucius (fifth or sixth century B.C.) also believed that the reciprocal relationships of the golden rule were the way to unite most harmoniously the diversities of human interests. His teaching was similar to the Old Testament's "Do not do to others what you would not wish others to do to you."[78]

Corresponding to the practical wisdom of Confucius is the mystical wisdom of the Tao Te Ching, attributed to Lao Tzu (seventh century B.C.). It expresses the Golden Rule in positive terms, as in the New Testament's "Do good to them that hate you."[79] For Lao Tzu, the power of limits becomes the key to harmonious human behavior, revealing the unity of minor and major human affairs: "Governing a large state is like boiling a small fish." And this recommendation to beware of excess: "He who tiptoes cannot stand; he who strides cannot walk; . . . he who boasts cannot endure." The natural world which unites large and small in the "greatness of little things" is constantly held up by the Tao Te Ching as the best guide to the art of living: "In the world there is nothing more submissive and weak than water. Yet for attacking that which is hard and strong nothing can surpass it."[80]

Possibly the oldest book of wisdom to influence the arts of living in China is the *Book of Changes* or *I Ching*, which originated, according to tradition, about 4000 years ago. It is based on the recognition that the ever-changing diversities of existence have an underlying unity of order, in which everything is related to everything else. The foundation of this order is the dinergic unity of the dark (yin) and the light (yang) principles, the former represented by a broken line — —, the latter by a solid line ——. These lines are grouped by threes into eight *trigrams,* which are put together in all possible combinations to form sixty-four *hexagrams,* each consisting of six lines and each symbolizing a different basic life situation.

Over the centuries, the sages of China have added commentaries to each of these signs. The book has been consulted throughout the ages as an oracle, to gain advice as to the proper attitude to be taken in any particular circumstance. The advice is usually couched in picturesque, mytho-poetic language, with images often taken from nature. The wisdom of nature is

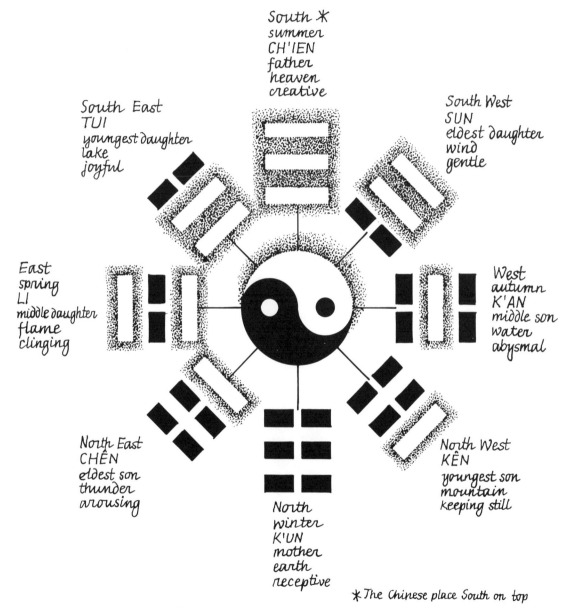

South ✳
summer
CH'IEN
father
heaven
creative

South East
TUI
youngest daughter
lake
joyful

South West
SUN
eldest daughter
wind
gentle

East
spring
LI
middle daughter
flame
clinging

West
autumn
K'AN
middle son
water
abysmal

North East
CHÊN
eldest son
thunder
arousing

North West
KÊN
youngest son
mountain
keeping still

North
winter
K'UN
mother
earth
receptive

✳ The Chinese place South on top

Fig. 169. Primal arrangement of *I Ching's* trigrams.

always held up as the best example of how to live. The so-called *primal arrangement* of the eight basic trigrams, with their main meanings, is shown in figure 169, grouped around the central yin-yang symbol, called *Tai Chi*, expressing the unity of diverse light and dark elements.[81]

The arts of living in the West have been shaped by knowledge rather than by wisdom, as we shall see from the examples that follow. But, indeed, the diversities of Eastern wisdom and Western knowledge seem to have common roots in the same basic pattern-forming processes that create the harmonies of nature and art.

In an eighteenth-century letter from an exiled Hungarian freedom fighter, Kelemen Mikes, the following anecdote appears, about a bishop whose own religious belief did not prevent him from becoming united with another human being, even one who expressed his faith in quite contrary terms to the bishop's.

This bishop, landing on a lonely island, found there an ignorant hermit in fervent prayer, repeating the words "Cursed be the Lord." The bishop sternly reprimanded him and corrected his prayer: "Blessed be the Lord!" Then he re-embarked and his boat soon lifted anchor. The sails were already catching the wind when a man was seen running after the boat. He caught up with it and climbed on board, to the amazement of the sailors, who thought he had run on top of the waves! It was the hermit, in tears, for he had forgotten the right words. The bishop, himself deeply moved, lifted the hermit from his knees, embraced and kissed him saying "Pray as you always prayed."[82]

If we look for unity in diversity in the two major political and economic powers today, the U.S.S.R. and the U.S.A., we find them both lacking such unity in quite opposite ways. The U.S.S.R. strives for monolithic unity and accomplishes it, but at the cost of stifling and condemning all diversity. The U.S.A., on the contrary, strives for complete freedom of diversities, which it achieves to a large extent, but at the cost of unity of purpose. The reciprocal relationship of neighbors, uniting dinergic diversities, is sadly missing in both cases, with a resultant lack of social harmony and strength.

The principle of the relatedness of neighbors was, however, part of the wisdom that founded this country. Thomas Jefferson realized that all human beings "are endowed by their creator" with equal rights to "life, liberty and the pursuit of happiness," because none are born "with saddles on their backs, nor a favored few booted and spurred. . . ."[83] Such was the Western world's contribution to the advancement of harmonious living in the eighteenth century. Today, our culture's contribution to human knowledge is largely in the realm of high technology, yet even here the principle of dinergy applies.

Knowledge of aerodynamics, of the structural strength of materials, of the physics and chemistry of engines, and of business skill would seem the most obvious considerations in the design of an aircraft. The design of the Boeing 747, however, also reveals harmonious proportions strikingly similar to those we have seen in natural forms and works of art.

As figure 170 shows, the entire plan **C** of this aircraft fits into two pairs of reciprocal golden rectangles, which meet along the centerline of the fuselage; the length of the larger ones corresponding to the front of the plane (from nose to end of antennae, at the wingtips) while the smaller ones encompass the tail end. The sideview **B** shows that the entire body is contained in five golden rectangles plus a reciprocal; the smaller one of these fits the nose, four of the larger ones embrace the rest of the fuselage, including wings and tail, while the fifth contains the rudder. Frontview **A** is contained in two golden rectangles. The wheel base is in golden proportion to the height of the plane (from the ground to the top of the rudder). (See wave diagram between **A** and **B** as well as graphs **A** and **B**.)

The proportions of each wing **D** recall the winged seeds of the maple tree and the wings of flies. Each wing is contained within a pair of equal golden rectangles, as the frontview is. The wave diagram of the wing shows how even the position of the engines along the leading edge and the break in the trailing edge of the wing share the very same relationships. Wave diagrams along the length and width of plan **C** show similarly shared relationships between the width of the fuselage, the breakline of the wings and the full wingspan, and between the front of the fuselage, the wingspan, and the tail end. The graph shows how these proportions approximate the root harmonies of music. While this plane is the product of technological, scientific and business knowledge, it is also a work of art with a beauty all its own. Such beauty is more than pleasing form: it is also wholeness and strength.

These random examples were meant to suggest how wisdom and knowledge have contributed in their diverse ways to the arts of living in East and West, and how our Western knowledge needs to be complemented at times by elements of Eastern wisdom, so that harmonious proportions may fully unfold not only in our airplane designs, but also in the patterns of our daily living. From the unity of diversities as wide apart as Buddhism and aircraft design, patterns of wholeness have emerged. A more detailed inquiry into the nature of wholeness as related to the arts of living shall be our final topic.

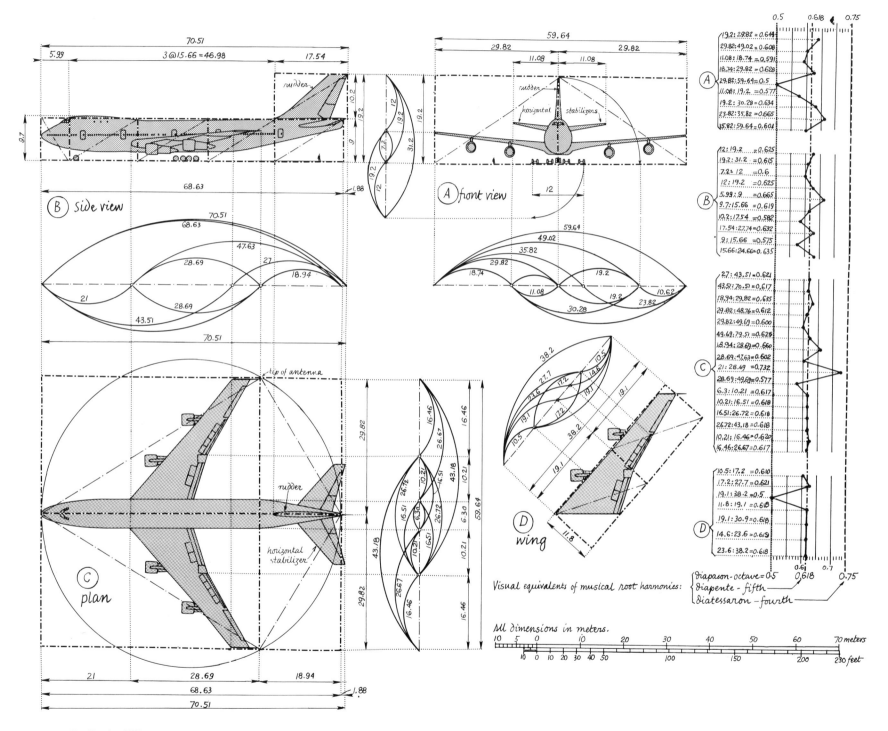

Fig. 170. Boeing 747.

Fig. 171. Mandala pattern of a thistle flower.

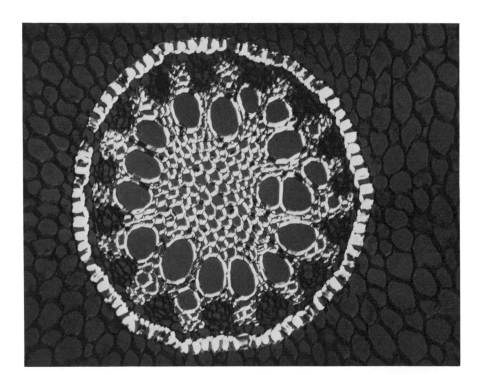

Fig. 172. Mandala pattern in the stem of a lily; enlarged 120 times.

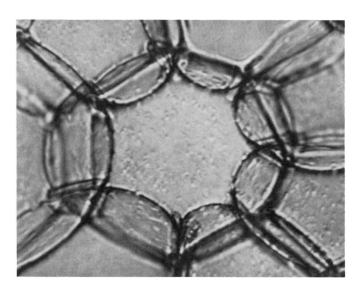

Fig. 173. Mandala pattern in the stem of a poppy; enlarged 1000 times.

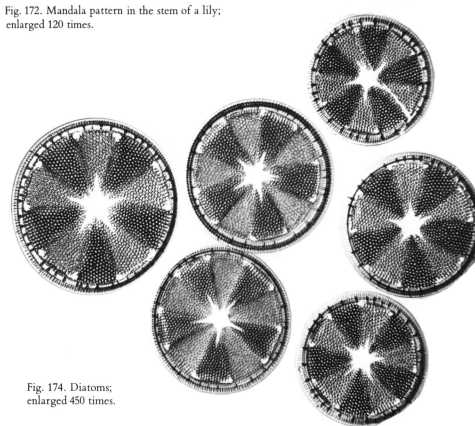

Fig. 174. Diatoms; enlarged 450 times.

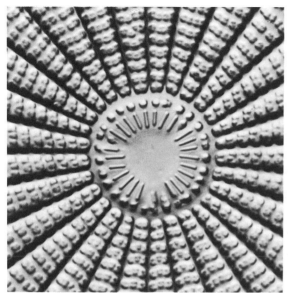

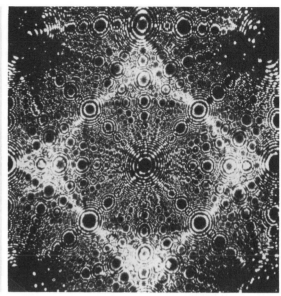

Fig. 175. Detail of diatom center; enlarged 2000 times.

Fig. 176. Mandala pattern created in liquid by harmonic vibrations.

Fig. 177. Square and circle in atomic pattern at tip of platinum needle; enlarged 750,000 times.

Whole, hell and holy

The art as well as the wisdom and the knowledge of East and West alike testify that there exists a deep-rooted unity below the many surface diversities of this world. This unity manifests itself in simple proportional relationships that create patterns of harmonious wholeness out of the vast and often dinergically opposed diversities in nature, in the arts, and, at times, in the arts of living.

The etymology of the word *whole* comes from the Indo-European root *kailo,* from which also come *hale, health, hallowed,* and *holy.* The word *hell,* though of similar sound, is of quite different origin, coming from the root *kel.* From this root comes *calamity, clamor, concealment,* and *calumny.*

Whole, hale, and holy on one side and hell on the other express the ultimate potentialities in human life: good and evil. These ultimate powers, the divine and the demonic, are usually thought of as supernatural and are looked upon with awe. The word *numinous* has been used to describe such feelings, a word that comes from the Latin *numen,* meaning divinity or spirit, from a root verb meaning a divine nod or command of compelling power.

But numinous feelings are not the exclusive domain of the supernatural. Quite to the contrary, the marvels of nature awaken in us similarly numinous feelings, particularly the wholeness patterns of nature, such as the thistleflower's structural pattern, (fig. 171) the cells within the stem of a lily (fig. 172) or a poppy (fig. 173), or the intricate single cells of tiny water plants called diatoms. (figs. 174, 175) Inorganic patterns, too, are awesome in their wholeness, as for instance the sound frequencies in figure 176.

In the Middle Ages, holy men argued about the number of angels that could dance on the end of a sword. Today we see in photographs of atomic patterns, taken with the help of the ion microscope (fig. 177) that not only a few angels but entire worlds dance in the tip of a needle!

Plan

Vestibule: The Opportunists ....

1: The Virtuous Pagans ....

2: The Carnal ....

3: The Gluttonous ....

4: The Hoarders & Wasters ....

5: The Wrathful & Sullen ....

6: The Heretics ....

7: The Violent & Bestial ....

8: The Fraudulent (Evil Ditches) ....

9: The Frozen Lake of The Loveless (Cocytus) ....

Patterns of wholeness have been created throughout the ages in all parts of the world, in crafts, arts, and architecture, giving tangible form to the itangible order that unites the diversities of this world. The East developed such patterns, which in Sanskrit are called *mandalas,* as aids to meditation and worship. The plan of the great Borobudur stupa is such a mandala pattern. Poets have also chosen mandala patterns at times to express human wholeness, the agony of being without it, and the struggle to achieve it. *The Divine Comedy,* written at the beginning of the thirteenth century by Dante Alighieri, presents a journey through the patterns of human destiny in accordance with the conceptions of medieval Christianity. The three parts of the poem, Hell, Purgatory and Paradise, are each conceived in the shape of immense mandalas.

Hell, as Dante describes it, is a tremendous funnel-shaped pit, like the crater of a volcano, into which the poet descends, guided by the

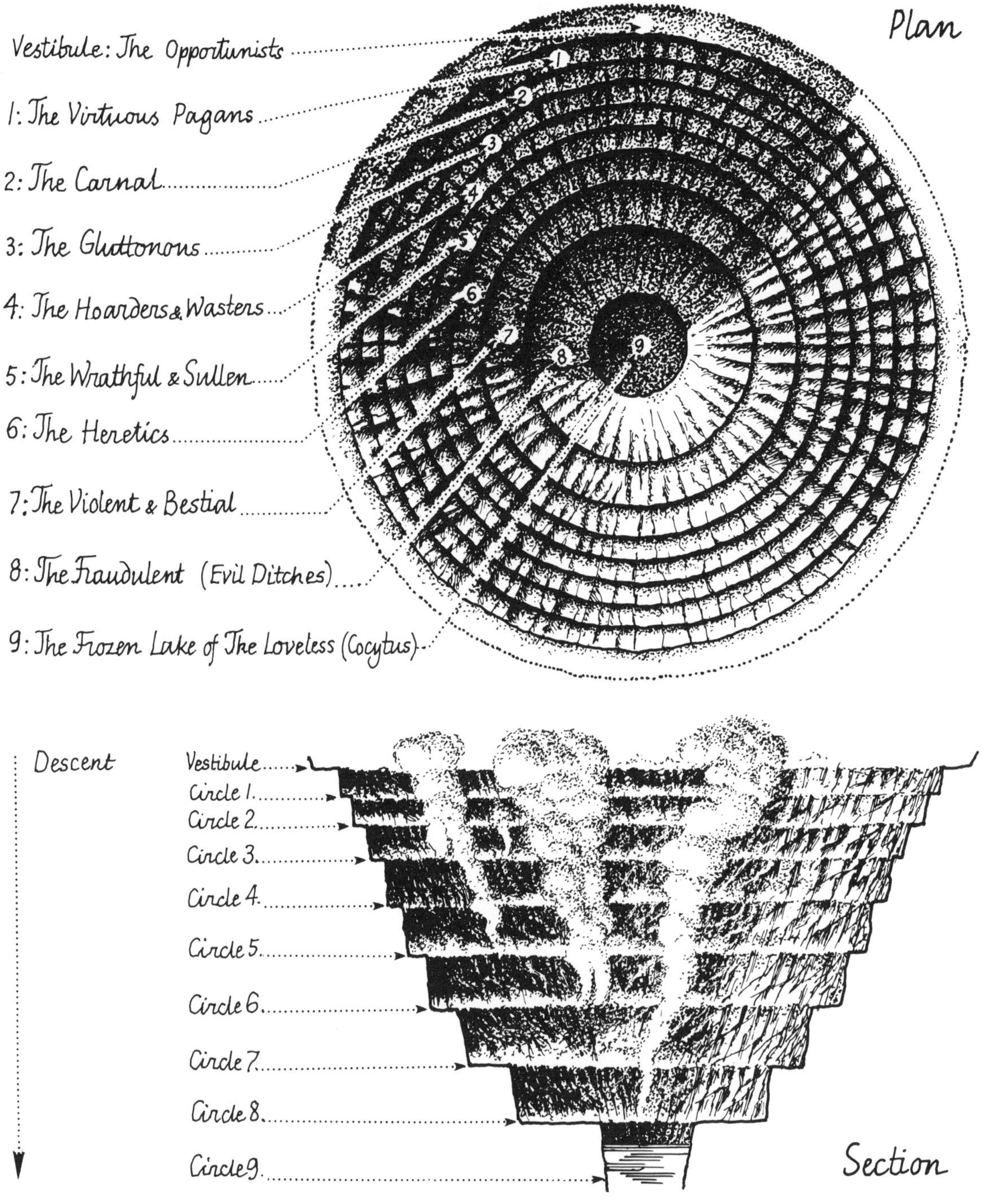

Descent

Vestibule ....
Circle 1 ....
Circle 2 ....
Circle 3 ....
Circle 4 ....
Circle 5 ....
Circle 6 ....
Circle 7 ....
Circle 8 ....
Circle 9 ....

Section

Fig. 178. Mandala pattern of Dante's Hell.

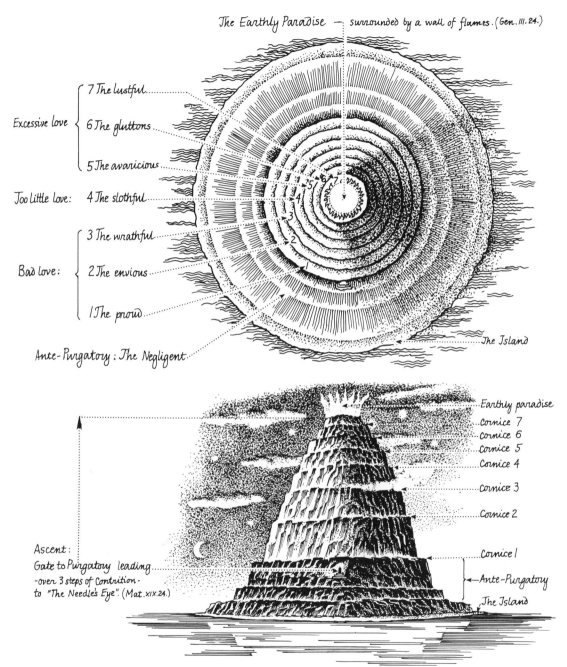

The Earthly Paradise — surrounded by a wall of flames. (Gen. III. 24.)

Excessive love
{
7 The lustful
6 The gluttons
5 The avaricious
}

Too little love: 4 The slothful

Bad love:
{
3 The wrathful
2 The envious
1 The proud
}

Ante-Purgatory: The Negligent

The Island

Earthly paradise
Cornice 7
Cornice 6
Cornice 5
Cornice 4
Cornice 3
Cornice 2
Cornice 1
Ante-Purgatory
The Island

Ascent:
Gate to Purgatory leading
-over 3 steps of Contrition-
to "The Needle's Eye". (Mat. XIX. 24.)

Fig. 179. Mandala and diagram of Dante's Purgatory.

Latin poet Virgil, whom Dante refers to as his ancestor. They move in ever narrowing circles, deeper and deeper, shuddering at the sight of the tormented who in their lives had cut themselves off from human relationships by their excesses of incontinence, violence and fraud. At the bottom of Hell's hole, at the ninth circle, are those who had treacherously denied all bonds of love and compassion. Here they are found frozen in the icy lake of lovelessness. (fig. 178)

Purgatory (fig. 179) is the inverse of Hell's pit: a mountain shaped like a giant pinecone, in plan likewise a mandala, which the poets ascend with great difficulty. Along craggy, rock-strewn spiraling pathways they find those who even after death suffer from various forms of egoism: pride, envy, sloth, avarice, and so on. (Vices which represent, in John Ciardi's words, deviations of love: "bad love, too little love, and immoderate love.")

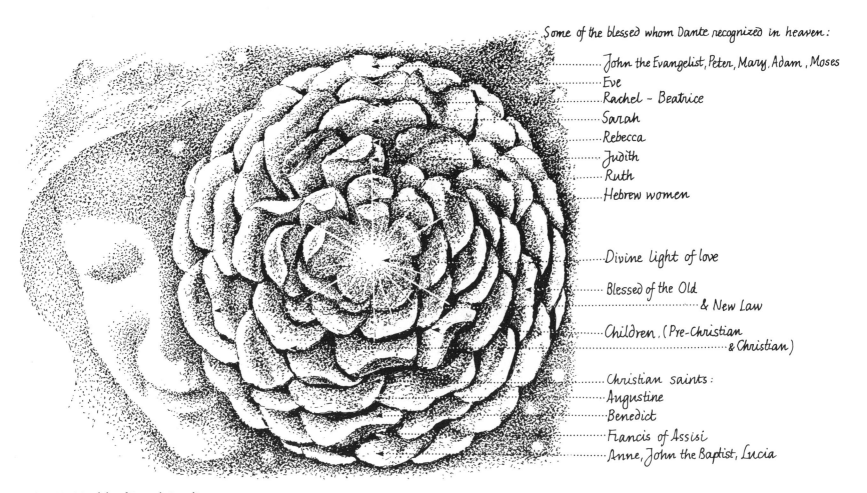

Some of the blessed whom Dante recognized in heaven:

John the Evangelist, Peter, Mary, Adam, Moses
Eve
Rachel - Beatrice
Sarah
Rebecca
Judith
Ruth
Hebrew women

Divine light of love

Blessed of the Old
& New Law

Children. (Pre-Christian
& Christian)

Christian saints:
Augustine
Benedict
Francis of Assisi
Anne, John the Baptist, Lucia

Fig. 180. Mandala of Dante's Paradise.

From the top of Mount Purgatory, Dante ascends into Paradise, where he is guided by Beatrice, the great love of his life. Paradise reveals itself as the ultimate vision of harmony, stretching through the heavens like an immense rose, (fig 180) bathed in radiant light. The poem ends with the following words:

Here my powers rest from their high fantasy
but already I could feel my being turned—
instinct and intellect balanced equally
as in a wheel whose motion nothing jars—
by the love that moves the Sun and the other Stars.[84]

The outward pilgrimage of the *Divine Comedy* is really the inward journey of all of us. We all fail to share ourselves in love. We all know the frozen lake of lovelessness. We all scramble over the craggy rocks of egoism, and if we are fortunate we may eventually see the light of "love that moves the Sun and the other Stars."

A modern counterpart to this medieval mandala is the "Song of the Cosmos" written by the Hungarian poet Attila Jozsef (1905-1937) when he was only eighteen. It is also an outward journey reflecting an inward one: a trip into the modern world, a world still in need of shared human wholeness. Outwardly, it is the trip of a lonely planet, orbiting in icy outer space, "seared by the winds of void." Inwardly, it is the journey of the young poet, who grew up in utmost poverty, whose father left the family in search of work, whose mother tried to support her three small children on meager wages earned by washing and cleaning for others.

The chaos and lack of wholeness in the poet's life is mirrored by the stream of unrelated imagery that makes up the poem— violent flashes, nostalgic scenes of purity and beauty, all expressing a desperate yearning for personal and social harmony. The verseform, however, relates everything in a wholeness-pattern according to the rules of the *Sonetti Corona* of the Italian Renaissance.

The "Song of the Cosmos" consists of fourteen sonnets. The last line of each sonnet is the same as the first line of the following one, until the last line of the last sonnet becomes the first line of the first sonnet. Finally, all these first and last lines comprise a Master Sonnet which conveys the succinct message of all fourteen sonnets combined. A kind of electrical circuit of energy is generated by this dinergy: the tightly ordered structure and the

Fig. 181. Mandala pattern of Jozsef's "Song of the Cosmos."

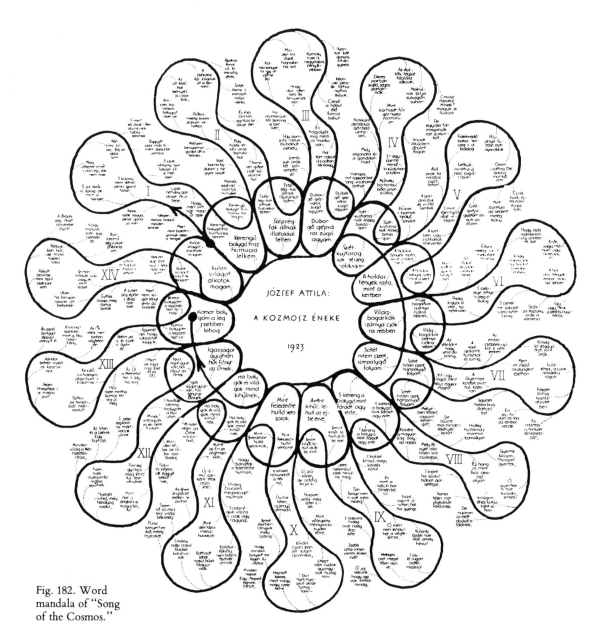

Fig. 182. Word
mandala of "Song
of the Cosmos."

freedom of diverse, anguished images. Here is a translation of the Master Sonnet:

I am a world all by myself alone,
my soul is fresh soil of a spinning planet's sphere.
Trees of beauty, full of fragrance flourish here;
my brain: a city, filled with motor-drone.

Moonlight patterns weave black with silver
in my nocturnal grove, like drunken men.
Worlds wing by like bugs, to mate-dance in the glen
over my dark faith: my sacred river.

My planet spins, as my worn brain does at night;
it cools off and falls, disappearing from light,
like lines of poems forgotten in my youth.

If all the worlds and all planets shall turn cold
one cool light shall flare up in the void, most bold,
kindled by the blaze of my lone planet's truth.[85]

Figure 181 is a pattern of the word-mandala of the "Song of the Cosmos" prepared by this author in 1945 with Dr. Janos Lotz.[86] We used traditional Hungarian flower patterns to represent lines of the sonnets. The major loops correspond to the fourteen sonnets, the winding of the vine indicates the stanzas of each sonnet—two quatrains and two terzets. The central garland represents the Master Sonnet. Figure 182 shows the complete text in Hungarian, with looping lines (the vine in the previous drawing) indicating the sequence.

The drawings present a visual equivalent of the harmonious structure of this poem. The "Song of the Cosmos" has the pattern of a life-saving ring, constructed from the realities of the modern world, to save us from drowning in a sea of lovelessness. The intensity of the struggle for love and harmony gives both the *Divine Comedy* and the "Song of the Cosmos" their awe-inspiring, numinous power. Both poems show us that there is a limitless order to our existence, an order of wholeness as unfathomable as the order of the cosmos, and as dreadful when violated as the order of the atom.

The ancient Hebrew wisdom of the Kabbalah holds that we are split beings, living in a split world. It teaches that our task in life is to restore to wholeness as many fragments as we encounter along the path of our life. This is the art of being human. The Hebrew word *shalom* preserves this wisdom in a nutshell: it means not only "peace," but also "wholeness."

We are all adolescents at the threshold of a new age of maturity. Perhaps one of the ways leading to the wisdom of maturity is the pathway of proportions, of shared limitations, which we need to find beneath the weeds and underbrush that have overgrown it. Proportions are shared limitations. As relationships they teach us the mana of sharing. As limitations they open doors toward the limitless. This is the power of limits.

The power of limits is the force behind creation. When two pebbles are dropped into the water some distance from each other, circular patterns of the parent waves join to create ellipses that grow wider and wider, until—beyond the confines of this picture—they also become circles. (fig. 183) Meanwhile, the first circles are becoming transformed: from closed circles, centered upon themselves, they are growing into parabolic arches, reaching out beyond themselves toward the infinite.

Is this merely a pattern of pebbles, of vibration, or is it also a metaphor for love, for the power of shared limits, and for the creative act itself?

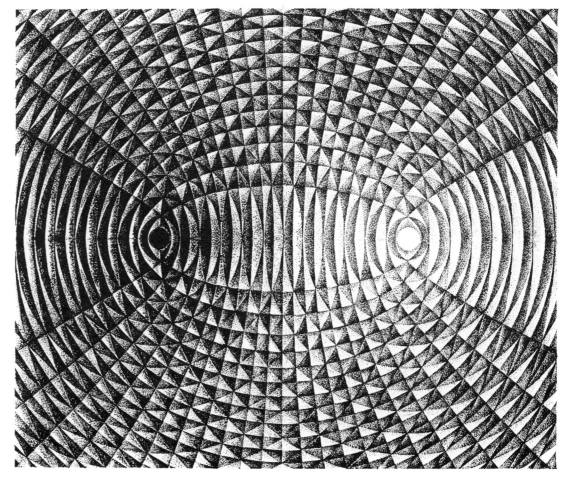

Fig. 183. The creative act.

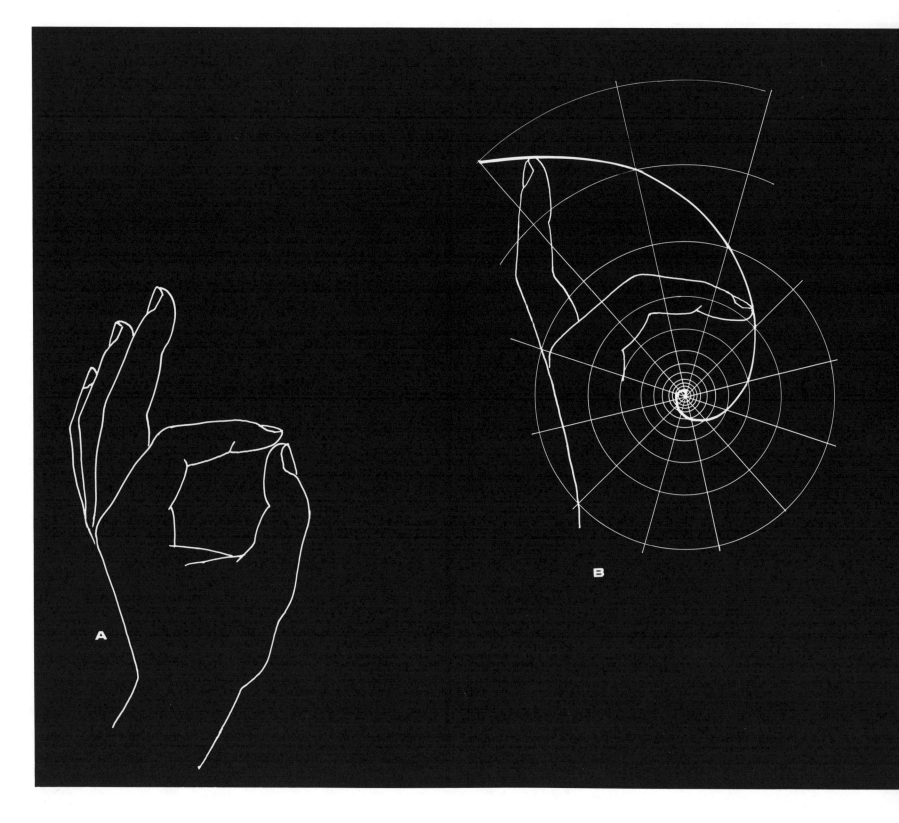

A

B

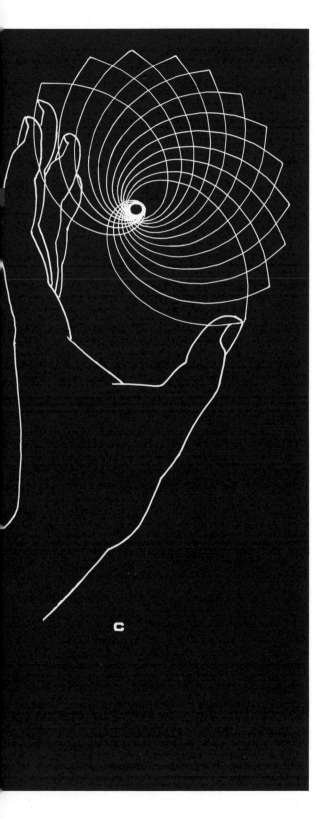

When we share our own limitations with the limitations of others, as we do in the golden relations of neighbors, we complement our own and others' shortcomings, creating thereby living harmony in the art of life, comparable to the harmonies created in music, dance, marble, wood and clay. It is possible to live in this way because the proportions of reciprocal sharing, nature's own golden proportions, are built into our own nature, into our bodies and minds which are, after all, part of nature. The basic pattern-forming processes of nature, which have shaped the human hand and mind, *can* continue to guide whatever the hand and mind are shaping, when the hand and mind are true to nature.

Thus the best human creations are ageless and even holy, like a freshly opened flower. Looking back to where we started, with Buddha holding up a flower (fig. 184), we see his hand first in the gesture of teaching (**A**). As it unfolds, the index finger moves along the same kind of logarithmic spiral that unfolds in the myriad forms of nature (**B**). The combined movements of all five fingers (**C**) create a picture of a thousand-petalled lotus, symbol of final attainment and wholeness. But this hand is not only the hand of the Buddha; it is the hand of every human being.

Fig. 184. The hand unfolding.

Appendices

Appendix I. Comparative Table of Human Proportions

The graph of Appendix I shows that those proportions which refer to directly connected neighboring bones (items 6-24) show greater unity in women and men of different sizes than the rest of the proportions investigated. Greatest deviation is dimension *a* (the chin's vertical distance from the line of the shoulders) which is proportionately longer in small men than in all others. The data having been gathered for military purposes, one might speculate that small men perhaps hold up their chins higher than others do.

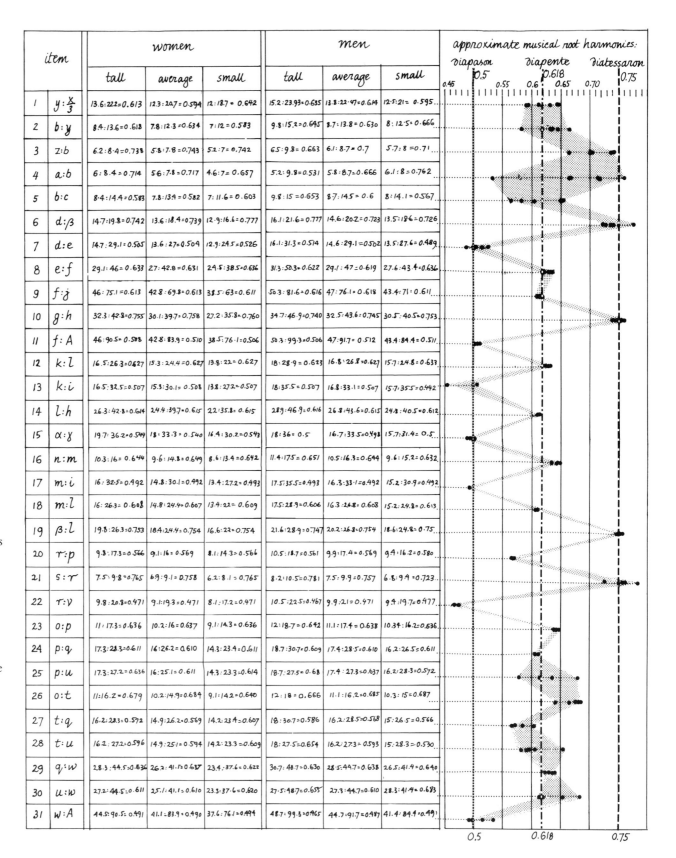

item	women tall	women average	women small	men tall	men average	men small
1 $y:\frac{x}{3}$	13.6:22.2=0.613	12.3:20.7=0.594	12:18.7=0.642	15.2:23.93=0.635	13.8:22.47=0.614	12.5:21=0.595
2 $b:y$	8.4:13.6=0.618	7.8:12.3=0.634	7:12=0.583	9.8:15.2=0.645	8.7:13.8=0.630	8:12.5=0.666
3 $z:b$	6.2:8.4=0.738	5.8:7.8=0.743	5.2:7=0.742	6.5:9.8=0.663	6.1:8.7=0.7	5.7:8=0.71
4 $a:b$	6:8.4=0.714	5.6:7.8=0.717	4.6:7=0.657	5.2:9.8=0.531	5.8:8.7=0.666	6.1:8=0.762
5 $b:c$	8.4:14.4=0.583	7.8:13.4=0.582	7:11.6=0.603	9.8:15=0.653	8.7:14.5=0.6	8:14.1=0.567
6 $d:\beta$	14.7:19.8=0.742	13.6:18.4=0.739	12.9:16.6=0.777	16.1:21.6=0.777	14.6:20.2=0.723	13.5:18.6=0.726
7 $d:e$	14.7:29.1=0.505	13.6:27=0.504	12.9:24.5=0.526	16.1:31.3=0.514	14.6:29.1=0.502	13.5:27.6=0.489
8 $e:f$	29.1:46=0.633	27:42.8=0.631	24.5:38.5=0.636	31.3:50.3=0.622	29.1:47=0.619	27.6:43.4=0.636
9 $f:g$	46:75.1=0.613	42.8:69.8=0.613	38.5:63=0.611	50.3:81.6=0.616	47:76.1=0.618	43.4:71=0.611
10 $g:h$	32.3:42.8=0.755	30.1:39.7=0.758	27.2:35.8=0.760	34.7:46.9=0.740	32.5:43.6=0.745	30.5:40.5=0.753
11 $f:A$	46:90.5=0.508	42.8:83.9=0.510	38.5:76.1=0.506	50.3:99.3=0.506	47:91.7=0.512	43.4:84.4=0.511
12 $k:l$	16.5:26.3=0.627	15.3:24.4=0.627	13.8:22=0.627	18:28.9=0.623	16.8:26.8=0.627	15.7:24.8=0.633
13 $k:i$	16.5:32.5=0.507	15.3:30.1=0.508	13.8:27.2=0.507	18:35.5=0.507	16.8:33.1=0.507	15.7:35.5=0.442
14 $l:h$	26.3:42.8=0.614	24.4:39.7=0.615	22:35.8=0.615	28.9:46.9=0.616	26.8:43.6=0.615	24.8:40.5=0.612
15 $\alpha:\gamma$	19.7:36.2=0.544	18:33.3=0.540	16.4:30.2=0.543	18:36=0.5	16.7:33.5=0.498	15.7:31.4=0.5
16 $n:m$	10.3:16=0.644	9.6:14.8=0.649	8.6:13.4=0.642	11.4:17.5=0.651	10.5:16.3=0.644	9.6:15.2=0.632
17 $m:i$	16:32.5=0.492	14.8:30.1=0.492	13.4:27.2=0.493	17.5:35.5=0.493	16.3:33.1=0.492	15.2:30.9=0.492
18 $m:l$	16:26.3=0.608	14.8:24.4=0.607	13.4:22=0.609	17.5:28.9=0.606	16.3:26.8=0.608	15.2:24.8=0.613
19 $\beta:l$	19.8:26.3=0.753	18.4:24.4=0.754	16.6:22=0.754	21.6:28.9=0.747	20.2:26.8=0.754	18.6:24.8=0.75
20 $r:p$	9.8:17.3=0.566	9.1:16=0.569	8.1:14.3=0.566	10.5:18.7=0.561	9.9:17.4=0.569	9.4:16.2=0.580
21 $s:r$	7.5:9.8=0.765	6.9:9.1=0.758	6.2:8.1=0.765	8.2:10.5=0.781	7.5:9.9=0.757	6.8:9.4=0.723
22 $r:v$	9.8:20.8=0.471	9.1:19.3=0.471	8.1:17.2=0.471	10.5:22.5=0.467	9.9:21=0.471	9.4:19.7=0.477
23 $o:p$	11:17.3=0.636	10.2:16=0.637	9.1:14.3=0.636	12:18.7=0.642	11.1:17.4=0.638	10.34:16.2=0.636
24 $p:q$	17.3:28.3=0.611	16:26.2=0.610	14.3:23.4=0.611	18.7:30.7=0.609	17.4:28.5=0.610	16.2:26.5=0.611
25 $p:u$	17.3:27.2=0.636	16:25.1=0.611	14.3:23.3=0.614	18.7:27.5=0.68	17.4:27.3=0.637	16.2:28.3=0.572
26 $o:t$	11:16.2=0.679	10.2:14.9=0.684	9.1:14.2=0.640	12:18=0.666	11.1:16.2=0.685	10.3:15=0.687
27 $t:q$	16.2:28.3=0.572	14.9:26.2=0.569	14.2:23.4=0.607	18:30.7=0.586	16.2:28.5=0.568	15:26.5=0.566
28 $t:u$	16.2:27.2=0.596	14.9:25.1=0.594	14.2:23.3=0.609	18:27.5=0.654	16.2:27.3=0.593	15:28.3=0.530
29 $q:w$	28.3:44.5=0.636	26.2:41.1=0.637	23.4:37.6=0.622	30.7:48.7=0.630	28.5:44.7=0.638	26.5:41.4=0.640
30 $u:w$	27.2:44.5=0.611	25.1:41.1=0.610	23.3:37.6=0.620	27.5:48.7=0.655	27.3:44.7=0.610	28.3:41.4=0.683
31 $w:A$	44.5:90.5=0.491	41.1:83.9=0.490	37.6:76.1=0.494	48.7:99.3=0.965	44.7:91.7=0.487	41.4:84.4=0.491

approximate musical root harmonies: diapason · diapente 0.618 · diatessaron — 0.5   0.618   0.75

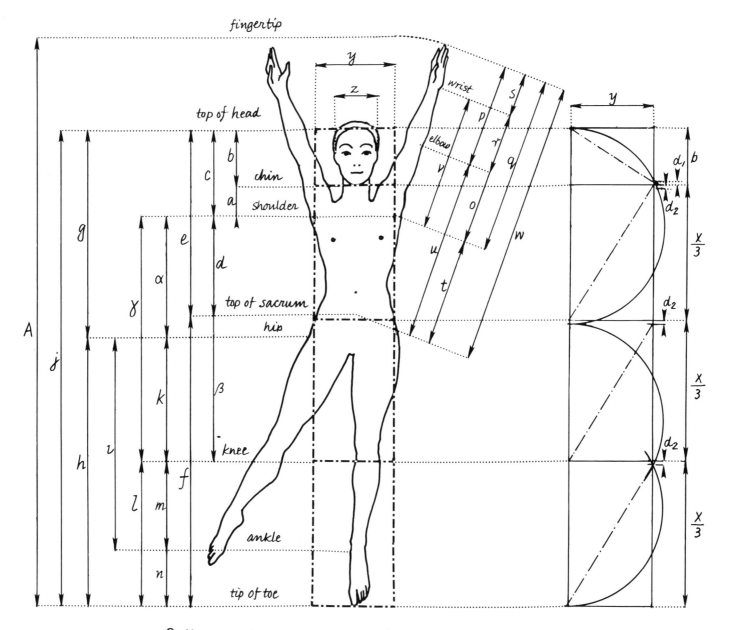

Differences between theoretical and actual golden proportions – d_1 & d_2 – in inches

		d_1		d_2
in tall women	: 13.6 × 0.618 = 8.4 — 8.4–8.4 =	0	13.6 × 1.618 = 22 —— 22.2–22 =	0.2
in averagesize women:	12.3 × 0.618 = 7.6 — 7.8–7.6 =	0.2	12.3 × 1.618 = 19.9 — 20.7–19.9 =	0.8
in small women	: 12 × 0.618 = 7.42 — 7.42–7 =	0.42	12 × 1.618 = 19.42 — 19.42–18.7 =	0.72
in tall men	: 15.2 × 0.618 = 9.39 — 9.8–9.39 =	0.41	15.2 × 1.618 = 24.59 — 24.59–23.93 =	0.66
in average size men	: 13.8 × 0.618 = 8.53 — 8.53–8.7 =	0.17	13.8 × 1.618 = 22.33 — 22.47–22.33 =	0.14
in small men	: 12.5 × 0.618 = 7.72 — 8–7.72 =	0.28	12.5 × 1.618 = 20.23 — 21–20.23 =	0.77

Appendix II. Human Proportions.
For numerical values see Appendix I.

Notes

Notes to Chapter 1.

1. Professor Fechner's experiments in 1876 were extensive, showing that over 75% of a large number of randomly selected people prefer rectangles in the golden section's proportions to any other rectangles. See H.E. Huntley, *The Divine Proportion: A Study in Mathematical Beauty* (New York: Dover Publications, 1970), p.64. This book gives a detailed discussion of the golden section.

2. A few random samples, in inches:
Standard size paper: 8.5 × 11, two golden rectangles (5.5 ÷ 8.5 = 0.647)
Driver's license: 5.5 × 8.5 (5.5 ÷ 8.5 = 0.647)
Credit card: 5.3 × 8.6 (5.3 ÷ 8.6 = 0.616   =± ⅝ = 0.625
U.S. paper money: 2.6 × 6.1 (2.6 ÷ 6.1 = 0.426)   0.447 is the ratio of a
Bank checks: 2.85 × 5.9 (2.85 ÷ 5.9 = 0.483) $\sqrt{5}$ rectangle's two sides

3. C.H. Waddington, "The Modular Principal and Biological Form," *Module, Proportion, Symmetry, Rhythm,* ed. Gyorgy Kepes (New York: George Braziller, 1966), p.37.

4. William Blake, "Auguries of Innocence," line 1.

5. In the Middle Ages the pentagram was often carved into doors in the belief that it would avert evil spirits. In Geothe's *Faust,* Mephistopheles says he could pass by this symbol on Faust's door only because it was poorly carved—one leg of the pentagram was askew. Curiously, our own country's defenses are housed in a building designed along the lines of the Pythagorean star: the Pentagon.

6. See also Hugo Norden, "Proportions in Music," *Fibonacci Quarterly,* vol. 2, no. 3, Oct. 1964; Paul Larson, "The Golden Section in the Earliest Notated Music," *Fibonacci Quarterly,* vol. 16, no. 6, Dec. 1976; Erno Lendvai, *Bela Bartok: An Analysis of His Music,* trans. T. Ungar (London: Kahn & Averill, 1971).

Notes to Chapter 2.

7. Bill Holm, *Northwest Coast Indian Art: An Analysis of Form* (Seattle: University of Washington Press, 1965), p.75.

8. Jay Hambidge, *Dynamic Symmetry: The Greek Vase* (New Haven: Yale University Press, 1920). A good description and fair evaluation of Hambidge's work can be found in Walter Dorwin Teague, *Design This Day* (London: Studio Publications, 1947).

9. John Benjafield and Christine Davis, "The Golden Section and the Structure of Connotation," *The Journal of Aesthetics and Criticism,* Summer 1978, pp. 123–27.

10. Joseph Eppes Brown, ed., *The Sacred Pipe: Black Elk's Account of the Seven Rites of the Oglala Sioux* (New York: Viking Penguin, 1971) pp. 95–98.

11. F.H. Cushing, "A Study of Pueblo Pottery as Illustrative of Zuni Culture Growth," (Washington: Smithsonian Institution Bureau of Ethnology, 1882-83) pp. 510–11.

Notes to Chapter 3.

12. "Tapu" or "taboo" have many positive connotations in tribal cultures. See Kaj Birket-Smith, *Primitive Man and His Ways* (New York: New American Library, Mentor, 1963) p. 189; Thor Heyerdahl, *American Indians in the Pacific* (Oslo: Gyldendal Norsk Forlag, 1952), p. 144.

13. Frank Waters, *The Book of the Hopi* (New York: Random House, Ballantine Books, 1974) p. 29.

14. Robert Graves, *White Goddess: A Historical Grammar of Poetic Myth* (New York: Random House, Vintage Books, 1958) p. 98.

15. Loren Eiseley, *The Unexpected Universe* (New York: Harcourt, Brace, 1972) p. 288.

16. Julian Jaynes, *The Origin of Consciousness in the Breakdown of the Bicameral Mind* (Boston: Houghton Mifflin, 1977).

17. "The Canticle of Brother Sun," *The Writings of St. Francis of Assisi,* trans. Benan Fahy, (London: Burns and Oates) p. 130.

18. Albert Einstein, *The World As I See It* (New York: Philosophical Library, 1949) p. 5.

19. William James, *The Varieties of Religious Experience* (New York: Doubleday, Dolphin Books, 1978) p. 55.

20. Abraham H. Maslow, *Religions, Values, and Peak Experience* (New York: Viking Compass, 1970) pp. 59, x–xi.

Notes to Chapter 4.

21. Einstein, *The World As I See It,* p. 1.

22. Miguel de Unamuno, *The Tragic Sense of Life,* trans. J. Crawford Flitch, (New York: Dover Publications, 1921) p. 25.

23. Susan Langer, *Philosophy in a New Key: A Study in the Symbolism of Reason, Rite and Art* (Cambridge: Harvard University Press, 1957) p. 62.

24. *Seattle Post Intelligencer,* 11 June 1980, p. 16.

25. Herbert Kuhn, *The Rock Pictures of Europe* (Fair Lawn, N.J.: Essential Books, 1956) p. 99. The word for "sea" starts with "m" in many languages—Latin, *mare;* German, *meer;* French, *mer;* Hebrew, *majim.*

26. Franz Boas, *Primitive Art* (New York: Dover Publications, 1955) p. 92.

27. Ibid, p. 99

28. Gordon Childe, *Man Makes Himself* (New York: New American Library, Mentor, 1951), p. 150.

29. Micah 6:8.

30. Gerald S. Hawkins, *Stonehenge Decoded* (New York: Dell Publishing, Delta Books, 1966). See also Evan Hadingham, *Circles and Standing Stones* (New York: Doubleday Anchor, 1975).

31. Hawkins, *Stonehenge,* p. 47.

32. Livio Catullo Stecchini, "Notes on the Relation of Ancient Measures to

the Great Pyramid," in Peter S. Tompkins, *The Secrets of the Great Pyramid* (New York: Harper & Row, 1978) p. 368.

33. Tompkins, *Secrets of the Great Pyramid,* p. xiv.

34. Ibid., p. 157.

35. Manuel Amabilis Dominguez, *La Architectura Precolumbina En Mexico* (Mexico, D.F.: Editorial Orion, 1956).

36. Genesis 28:12.

37. Tompkins, *Secrets of the Great Pyramid,* p. 186–87.

38. James Jeans, *Science and Music* (New York: Dover Publications, 1968).

39. Jean Dauven, "Sur la correspondence entre sons musicaux et la couleurs," in P.W. Pickford, *Psychology and Visual Aesthetics* (London: Hutchinson Educational, 1972) pp. 81–85.

Notes to Chapter 5.

40. Jay Hambidge, *The Elements of Dynamic Symmetry* (New York: Dover Publications, 1967) p. 10; Matila Ghyka, *The Geometry of Art and Life* (New York: Dover Publications, 1978) pp. 94–97; Huntley, *Divine Proportion,* p. 165; and D'Arcy Thompson, *On Growth and Form,* abridged ed. (Cambridge: Cambridge University Press, 1968) ch. VI.

41. Peter Kropotkin, *Mutual Aid: A Factor of Evolution* (Boston: Porter Sargent Publishers, Extending Horizons Books, 1976) p. 25.

42. Karl von Frisch, *The Dance Language and Orientation of Bees,* trans. Leigh E. Chadwick (Cambridge: Harvard University Press, Belknap Press, 1967).

43. Kropotkin, *Mutual Aid,* p. 56.

44. Edward O. Wilson, *Sociobiology: The Abridged Edition* (Cambridge: Harvard University Press, Belknap Press, 1980) p. 220.

45. Kropotkin, *Mutual Aid,* p. 37.

46. Wilson, Sociobiology, p. 475.

47. Kropotkin, *Mutual Aid,* p. 30.

48. Ibid., p. xii.

49. Ashley Montagu, ed., *Learning Non-Aggression: The Experience of Non-Literate Societies* (New York: Oxford University Press, 1978) p. 6.

50. Sally Carrighar, "War Is Not in Our Genes," *Man and Aggression,* ed. Ashley Montagu (New York: Oxford University Press, 1973) p. 122.

51. Rene Dubos, "The Despairing Optimist," *The American Scholar,* vol. 40, no. 4, Autumn 1971.

52. W.C. Allee, "Where Angels Fear to Tread," *The Direction of Human Development,* ed. Ashley Montagu (New York: American Elsevier Publishers, Hawthorn Books, 1970) p. 25.

53. Colin M. Turnbull, "The Politics of Non-Aggression," *Learning Non-Aggression.*

54. Barbara Ward, *The Rich Nations and the Poor Nations* (New York: W.W. Norton & Co., 1962) p. 150.

Notes to Chapter 6.

55. George D. Birkhoff, *Aesthetic Measure* (Cambridge: Harvard University Press, 1933).

56. Huntley, *Divine Proportion,* p. 14.

57. Ibid., p. 33.

58. Theodore Andrea Cook, *The Curves of Life* (New York: Dover Publications, 1979).

59. Samuel Colman and C. Arthur Coan, *Nature's Harmonic Unity* (New York: G.P. Putnam's Sons, 1912) and *Proportional Form* (New York: G.P. Putnam's Sons, 1920).

60. *Montaigne's Essays and Selected Writings,* ed. Donald M. Frame (New York: St. Martin's Press, 1969) p. 413.

61. Thompson, *On Growth and Form,* pp. 268–325.

62. Ibid., p. 321.

63. Paul Weiss, "Beauty and the Beast: Life and the Rule of Order," *The Scientific Monthly,* Dec. 1955.

64. A good description of various harmonograph patterns as well as of harmonograph constructions can be found in H.M. Cundy and A.P. Rollett, *Mathematical Models* (New York: Oxford University Press, Clarendon Press, 1961).

65. Simone Weil, *Gravity and Grace* (New York: Octagon Books, 1979).

66. Donald J. Borror and Richard E. White, *A Field Guide to the Insects* (Boston: Houghton Mifflin Co., 1974).

67. Vitruvius Pollio, *The Ten Books on Architecture,* trans. Morris H. Morgan (New York: Dover Publications, 1960).

68. Joscelyn Godwin, *Robert Fludd* (Boulder, Colo.: Shambhala Publications, 1979).

69. Rudolf Wittkower, *Architectural Principles in the Age of Humanism* (London: Alec Tiranti Ltd., 1962) quoting William Hogarth, David Hume, Edmund Burke, and John Ruskin.

70. Ibid., p. 33.

71. Simone Weil, *The Need for Roots,* trans. Arthur Wills (New York: G.P. Putnam's Sons, 1952) pp. 285–87, 293.

Notes to Chapter 7.

72. Gyorgy Kepes, ed., *Module, Proportion, Symmetry, Rhythm* (New York: Braziller, 1966).

73. George E. Duckworth, *Structural Patterns and Proportions in Virgil's Aeneid* (Ann Arbor: University of Michigan Press, 1962).

74. Benjamin Rowland, Jr., "The Evolution of the Buddha Image," *The Asia House Gallery Publication* (New York: Arno Press, 1976).

75. Kakuzo Okakura, *The Book of Tea,* (N.Y.: Dodd, Mead, 1926) p. 6.

76. Harold G. Henderson, *An Introduction to Haiku,* (Garden City, N.Y.: Doubleday Anchor, 1958).

Notes to Chapter 8.

77. Erwin Schrödinger, *What Is Life?* (Cambridge: Cambridge University Press, 1962) p. 140. While fully recognizing the need for such a cultural blood transfusion, Schrödinger also urged that precautions be taken. He saw the

main difficulty in the Western eagerness to imitate the East instead of integrating its wisdom into our own knowledge.

78. Leviticus 9:18.

79. Matthew 5:44

80. Lao Tzu, *Tao Te Ching* (New York: Viking Penguin, 1964) pp. 121, 81, 140.

81. *The I Ching,* trans. Richard Wilhelm (Princeton, N.J.: Princeton University Press, 1967).

82. Kelemen Mikes participated in the Rakoczy rebellion against the Habsburgs and spent the last years of his life as a political refugee in Turkey. He tells the story in one of his letters home.

83. *Thomas Jefferson, Life and Selected Writings,* eds. Adrienne Koch and William Peden (New York: Random House, Modern Library, 1944) pp. 22, 729–30.

84. Dante Alighiere, *Paradiso,* trans. John Ciardi (New York: New American Library, Mentor, 1970).

85. Author's translation.

86. See John Lotz, *The Structure of the Sonnetti a Corona of Atilla Jozsef* (Stockholm: Almqvist & Wiksell, 1965).

Credits

Fig. 8: Bryant Brewer, Seattle; Fig. 12: After Karl Blossfeldt, *Urformen der Kunst;* Figs. 22, 24 & 25: After specimens at the Thomas Burke Memorial Washington State Museum, Seattle; Fig. 26: Museum für Deutsche Volkskunst, Berlin; Figs. 28 & 29: After Bill Holm, *Northwest Coast Indian Art;* Figs. 30 & 32: Seattle Art Museum; Fig. 33: After R. Higgins, *Minoan and Mycenaean Art;* Fig. 34: After J. Bazley and P. Jacobsthal, "Protogeometric Pottery," *American Journal of Archeology,* 1940; Fig. 35: After Higgins, *Minoan and Mycenaean Art;* Figs. 36, 37 & 38: After specimens at the Thomas Burke Memorial Washington State Museum, Seattle; Fig. 40: After Oswald White Bear Fredericks, *Book of the Hopi;* Fig. 41: After an illustration from the Victoria and Albert Museum, London; Fig. 42: Archaeological Museum, Heraklion; Fig. 43: Cartari, "Le imagini de i dei"; Fig. 44: After Lommel, *Prehistory and Primitive Man;* Fig. 46: National Museum of Antiquities of Scotland; Fig. 48: After Vesalius and de Burlet; Fig. 49: After James D. Watson, *The Double Helix;* Fig. 50: Dr. L. E. Roth, Dr. Y. Schigenaka, and Dr. D. J. Pihlaja; Fig. 52: Inside of a terracotta drinking cup from Greece, 6th century B.C. in the Museu Nazionale, Tarquinia; Fig. 53: Musee Guimet, Paris; Fig. 54: Radost Folk Ensemble, Seattle Theatre Arts, Seattle; photograph by Chris Bennion; Fig. 55: After Evan Hadringham, *Circles and Standing Stones;* Figs. 56, 57 & 58: After Franz Boas, *Primitive Art;* Fig. 59: The University Museum, The University of Pennsylvania; Fig. 61: After Donald Anderson, *The Art of Written Forms;* Fig. 63: British Museum, London; Fig. 64: National Museum, Tai-chung; Fig. 65: From E. L. Sukenik, *Treasures of the Hidden Scrolls* (Jerusalem, Bialik Institute); Fig. 69: After S. Giedion, *Eternal Presence;* Fig. 70: Ashmolean Museum; Figs. 71, 72 & 73: After Herbert Kühn, *The Rock Pictures of Europe;* Fig. 74: British Crown copyright, reproduced with the permission of the Controller of Her Brittanic Majesty's Stationery Office; Fig. 75: After Hadringham, *Circles and Standing Stones;* Fig. 76: British Crown copyright, reproduced with the permission of the Controller of Her Brittanic Majesty's Stationery Office; Fig. 77: Illustration from *Stonehenge Decoded* by Gerald S. Hawkins in collaboration with John B. White. Copyright © 1965 by Gerald S. Hawkins and John B. White. Reproduced by permission of Doubleday & Company, Inc; Fig. 78: After Hawkins, *Stonehenge Decoded;* Fig. 80: After Peter Tompkins, *Secrets of the Great Pyramid* and Bannister Fletcher, *A History of Architecture;* Fig. 84: After Jorge A. Acosta, *Teotihuacan Official Guide;* Fig. 85: After Doris Heyden and Paul Gendrup, *Precolumbian Architecture of MesoAmerica;* Fig. 86: After Ignazio Marquinas, *Estudio Architectonico Comparativo de los Monumentos Archeologicos de Mexico;* Fig. 87: The University Museum, the University of Pennsylvania; Fig. 88: After Leonard Wooley, *Ur Excavations;* Fig. 89: Gemälde Gallerie, Kunsthistorisches Museum, Vienna; Fig. 90: After Stecchini's reconstruction in Tompkins, *Secrets of the Great Pyramid;* Fig. 91: After Giedion, *Eternal Presence;* Figs. 92, 93: After Wooley, *Ur Excavations;* Fig. 95: After Sir James Jeans, *Science and Music;* Fig. 96: After Jean Dauven "Sur la correspondence entre les sons musicaux et la couleurs," *Couleurs* (Paris) no. 77, Sept. 1970; Fig. 97: Prescolite Co., Seattle; Fig. 108, 109 & 111: After drawings by D. R. Harriot in Bulletin 180, "Pacific Fishes of Canada" by J. L. Hart. Reproduced by permission of the Minister of Supply and Services, Canada; Fig. 110: After R. McN. Alexander, "Functional Design in Fishes"; Fig. 112: After a specimen at the Thomas Burke Memorial Washington State Museum, Seattle; Fig. 119: American Museum of Natural History; Fig. 120: After W. Ellenberger et al., *An Atlas of Animal Anatomy for Artists;* Fig. 121: After Darwin; Fig. 122: After Karl von Frisch, *The Dance Language and Orientation of Bees;* Fig. 123: Drawing by Sarah Landry in an account by Pilleri and Knuckay, *Zeitschrift für Tierpsychologie,* 26 (1); Fig. 124: After photographs by Alexandra Kruse and Eliot Porter; Figs. 125, 126: After a photograph by Grant Heilman; Fig. 129: Palomar Observatory, California Institute of Technology; Fig. 132: After D'Arcy Thompson, *On Growth and Form;* Fig. 135: After D. J. Borror and R. E. White, *A Field Guide to the Insects of America North of Mexico;* Fig. 136: After Borror et al., *An Introduction to the Study of Insects;* Fig. 137: After Borror and White, *A Field Guide to the Insects of America North of Mexico;* Fig. 139 & 141: After specimens from the collection of Donald Collins; Fig. 140: After a specimen at the American Museum of Natural History, New York; Fig. 143: Capodimonte Museum, Naples; Fig. 144: From *The Literary Remains of A. Dürer;* Fig. 145: From Joscelyn Godwin, *Robert Fludd;* Fig. 146: After Vesalius, *Anatomic Atlas;* Thomson, *A Handbook of Anatomy;* Gray, *Anatomy of the Human Body;* and Dreyfuss, *The Measure of Man;* Fig. 150: Photograph by Jim Cummins; Fig. 151: Museo Nazionale, Naples: Fig. 152: Museo Nazionale, Rome; Fig. 153: National Museum, Athens; Fig. 154: After Vitruvius, *Ten Books on Architecture;* Fig. 155: After J. Buhlmann, *Die Architektur des Classischen Altertums und der Renaissance* and Erik Lundberg, *Arkitekturens Formsprak;* Fig. 156: After Kohte, *Die Baukunst des Classischen Altertums* and Lundberg, *Arkitekturens Formsprak;* Fig. 157: After photographs and drawings in *Arco di Constantino,* Editorial Domus, Milan, and Fletcher, *A History of Architecture;* Fig 158: After Fletcher, *A History of Architecture;* Fig. 159: After Benjamin Rowland, *The Evolution of the Buddha Image;* Fig. 160: National Museum, Seoul, Korea; Fig. 161: After a photograph by Royal Dutch Indian Airways, and B. Namikawa et al., *Borobudur;* Fig. 162: After Lundberg, *Arkitekturens Formsprak,* and Oswald Siren, *Histoire des arts anciens de la Chine;* Fig. 163: After Okamoto, *The Zen Garden,* and Teiji Ito et al., *The Japanese Garden;* Fig. 164: After Arthur Drexler, *The Architecture of Japan,* and Werner Blaser, *Japanese Temples and Teahouses;* Fig. 165: After Drexler, *The Architecture of Japan;* Fig. 166: After Naomi Okawa, *Edo Architecture: Katsura and Nikko;* Fig. 167: After Drexler, *The Architecture of Japan;* Fig. 168: Seattle Art Museum; Fig. 169: After Richard Wilhelm's translation of the *I Ching;* Fig. 170: After drawings provided by the Boeing Company, Seattle; Figs. 172, 173, 174 & 175: Carl Strüwe, *Formen des Microcosmos* (Munich: Prestel Verlag, 1955). Reproduced with permission. Fig. 176: After Hans Jenny, *Cymatics;* Fig. 177: Erwin W. Muller, "The Field Ion Microscope," *The American Scientist,* vol. 49, no. 1, March 1961. Reproduced with permission. Figs. 178, 179 & 180: After translations of the *Divine Comedy* by Dorothy Sayers and John Ciardi; Figs. 181, 182: After John Lotz, *The Structure of the Sonnetti a Corona of Attila Jozsef* (Stockholm: Almqvist & Wiksell, 1965); Appendix I: Data from Henry Dreyfuss, *The Measure of Man.*

Index